nonlinear networks

John Elvey

nonlinear networks

VACLAV DOLEZAL

Department of Applied Mathematics and Statistics,
State University of New York at Stony Brook,
Stony Brook, N.Y. 11794, U.S.A.

ELSEVIER SCIENTIFIC PUBLISHING COMPANY
Amsterdam — New York — Oxford 1977

ELSEVIER SCIENTIFIC PUBLISHING COMPANY
335 Jan van Galenstraat
P.O. Box 211, Amsterdam, The Netherlands

Distributors for the United States and Canada:

ELSEVIER NORTH-HOLLAND INC.
52, Vanderbilt Avenue
New York, N.Y. 10017

Library of Congress Cataloging in Publication Data

Doležal, Václav.
Nonlinear networks.

Bibliography: p.
1. Electric networks. 2. Hilbert space. I. Title.
TK454.2.D65 1977 621.319'2 77-1464
ISBN 0-444-41571-8

Printed in The Netherlands

PREFACE

The aim of this book is to develop a general model of a nonlinear network governed by Kirchhoff laws which encompasses all electrical, mechanical and hydraulic networks encountered in engineering science.

With the boom in semiconductor technology of the past few decades, nonlinear networks have become the main objects of interest in today's network theory. Unfortunately, nonlinear networks lack many of the attractive properties that linear networks possess. For example, the "superposition principle" does not hold, "frequency domain analysis" is unapplicable, etc.

Another serious setback is the fact that, once we admit nonlinear elements as building blocks of a network, it is not immediately apparent whether the network is solvable for any applied excitation, i.e., whether or not there exists a current distribution corresponding to a given excitation. For example, we cannot connect a fluorescent tube directly to an AC source unless a choke is inserted in the circuit.

Similarly, even if a current distribution exists, there is no immediate guarantee that it is unique.

The facts display the sharp contrast between the properties of nonlinear and classical linear networks with constant elements. In the case of the latter networks, the situation is simple: for any excitation the current distribution exists and is unique, if and only if a certain determinant is nonzero (see [17]). Thus, apart from "pathological cases", every such network is well-behaved.

In past decades many papers appeared dealing with the analysis of nonlinear networks. However, in most instances a particular kind of network was considered. Scarcely any paper developed a sufficiently general network model.

The theory presented in this book attempts to fill this gap. Our main concern will be the so-called Hilbert network. We are led to the Hilbert network concept through noting the fact that various types of energies can be associated with actual networks, i.e., we are able to form certain products of network variables such as voltages, currents, etc. Also, dealing with a Hilbert space, we get an unexpected bonus: we can study infinite networks as well, and the results hold independently of whether the network is finite or infinite.

Roughly speaking, by a Hilbert network we understand an interconnection of finitely or countably many lumped elements, whose variables belong to some Hilbert space and obey Kirchhoff's laws. The network elements are assumed to be described by, in general, nonlinear multivalued operators. Hilbert networks will be studied in Chapters 2 - 5.

On the other hand, it is not absolutely essential to use a Hilbert space as the underlying space. Actually, we frequently encounter situations, in which the network variables are continuous functions, integrable functions, etc., which are not elements of any standard Hilbert space.

Fortunately, it transpires that some results on Hilbert networks carry over to this more general case in which the network variables are in some linear space not necessarily equipped with any topology. Of course, the corresponding theory cannot be developed so well as for Hilbert networks, since topology is missing in this setting. Thus, for example, we cannot consider infinite networks, and we are unable to get results that reflect energy relations.

This model, called an "algebraic network", will be discussed in Chapter 6.

To develop the theory of a Hilbert or algebraic network, it is convenient to study first a more general (and more primitive) structure - an abstract network. Specializing the abstract network in a certain way, we obtain a Hilbert network and the corresponding results. Thus, as a particular case we get a model of a nonlinear R, L, C - network, whose variables are in the space $L_2[0,\tau]$, $\tau \leq \infty$. Another specialization yields an algebraic network. The abstract network will be studied in Chapter 1.

Prerequisites needed for understanding the theory in this book are minimal: familiarity with the basics of functional analysis, Hilbert space theory and classical network theory is sufficient. For the convenience of the reader, most sections are augemented by problems which promote understanding of the material discussed and stimulate further thoughts. The book can thus serve as a text in advanced network theory.

It is a pleasure to acknowledge the support of the National Science Foundation under Grant PO 33568-X00 which made possible the research leading to the results in this book. I am indebted to my colleague, Professor A. H. Zemanian, for many valuable suggestions and discussions during the research. Finally, I am grateful to Mrs. C. Alberghine for her careful typing of the manuscript.

Stony Brook, New York V. DOLEZAL

TABLE OF CONTENTS

CHAPTER 1

ABSTRACT NETWORKS

1.1 Preliminaries

In this chapter we will introduce the concept of an abstract network and establish certain results which will be used in the following Chapter for studying Hilbert and algebraic networks.

To explain the basic idea of an abstract network, let us consider the classical example of a DC-network driven by voltage sources. Assume that our network $\hat{n}$ is described by a (finite) oriented graph G having branches $b_1,b_2,\ldots,b_{c_2}$ and nodes $n_1,n_2,\ldots,n_{c_1}$, and assume that branch b_j contains a resistor of resistance Z_j, $j = 1,2,\ldots,c_2$. Let d be the $c_2 \times c_1$ incidence matric of G, and let $Z' = \mathrm{diag}(Z_1,Z_2,\ldots,Z_{c_2})$. (Note that rows and columns of d correspond to branches and nodes of G, respectively. Many authors understand by "incidence matrix" our d^T.) If we insert a battery with voltage e_j in the branch b_j, and denote by i_j the current flowing through b_j, $j = 1,2,\ldots,c_2$, then, as is well known, the Kirchhoff laws governing our network can be formulated as follows:

$$K_1 : \qquad v^T \cdot (Z' \cdot i - e) = 0$$
$$\qquad\qquad \text{for each } v \in R^{c_2} \text{ with } d^T \cdot v = 0,$$

$$K_2 : \qquad d^T \cdot i = 0.$$

Here, R^{c_2} denotes the c_2-dimensional Euclidean space and $e = [e_1,e_2,\ldots,e_{c_2}]^T \in R^{c_2}$, $i = [i_1,i_2,\ldots,i_{c_2}]^T \in R^{c_2}$.

Note that K_2 is precisely the node law, and K_1 is equivalent to the loop law. Also, let us point out the fact that in this example, as well as in Chapters 1 - 4, we consider the simpler situation in which $\hat{\mathfrak{n}}$ is driven solely by EMF's, i.e., no independent current sources are present in $\hat{\mathfrak{n}}$. The more general case, when both kinds of sources are present, will be treated in Section 5.2, and we avoid it now only for the sake of clarity of the presentation.

If we put $N_o = \{x : x \in R^{c_2}, d^T \cdot x = 0\}$, then the above conditions K_1, K_2 can clearly be written in the form

K_1^o : $\qquad Z' \cdot i - e \in N_o^{\perp}$,

K_2^o : $\qquad i \in N_o$.

Note also that $R^{c_2} = N_o \oplus N_o^{\perp}$.

Obviously, the subspace N_o can be interpreted as the set of all possible current distributions in $\hat{\mathfrak{n}}$ that are determined solely by the structure of $\hat{\mathfrak{n}}$ (i.e., distributions obeying only the node law), and $N_o^{\perp}$ as the set of all possible vectors of total voltage drops in branches. Moreover, the matrix Z' can be regarded as a description of the electrical behavior of the network elements.

The conditions K_1^o, K_2^o already suggest how an abstract network may be defined. Indeed, we can replace R^{c_2} by an atstract linear space $\mathcal{L}$, the sets N_o, $N_o^{\perp}$ by subspaces N, M such that $\mathcal{L} = N \oplus M$, and the matrix Z' by some nonlinear (possibly multivalued) operator Z from a subset D of $\mathcal{L}$ into $\mathcal{L}$, and thus obtain an abstract network. Then, for a given "abstract voltage vector" $e \in \mathcal{L}$, we can define an "abstract current distribution" $i \in \mathcal{L}$ by means of relations analogous to K_1^o, K_2^o. Now, suppose that we derive a result on abstract networks, for example, conditions guaranteeing the existence and uniqueness of a current distribution $i \in \mathcal{L}$. Specifying then $\mathcal{L} = R^{c_2}$, $N = N_o$, $M = N_o^{\perp}$, we immediately obtain a corresponding result concerning a DC-network.

As we shall see, the concept of an abstract network proves to be much more powerful than what we could infer from the above trivial example of a DC-network; both Hilbert and algebraic networks can be reduced to an abstract network.

The abstract network will be defined in Section 1.2. Before going into this, we introduce several concepts and make some auxiliar observations.

Although we will work essentially with relations on sets, for our purposes it will be more convenient to adopt a slightly different approach.

Let X, Y be nonempty sets and let $\mathfrak{S}(Y)$ be the collection of all nonempty subsets of Y. A mapping $A : X \to \mathfrak{S}(Y)$ will be called a set mapping from X to Y.

If $D \subset Y$ is nonempty, we denote

$$(AD)^{\circ} = \bigcup_{x \in D} Ax. \tag{1.1.1}$$

If, in particular, A is a set mapping such tht Ax is a singleton for each $x \in X$, then A will be called an operator on X. Since we will not distinguish between a point $y \in Y$ and the singleton y , our operator is just an ordinary mapping from X into Y, and we have $(AD)^{\circ} = AD$.

Let A,B be set mappings from X to Y; if $Ax = Bx$ for every $x \in D$, we will write $A = B$ on D.

Definition: Let $A : X \to \mathfrak{S}(Y)$ be a set mapping, and let $D \subset X$, $D \neq \emptyset$; then the set mapping $A^{-} : (AD)^{\circ} \to \mathfrak{S}(D)$ defined by

$$A^{-}y = \{x : x \in D,\ y \in Ax\} \tag{1.1.2}$$

will be called the quasi-inverse of A on D.

If, in particular, A is an operator on D and A^{-} is also an operator on AD, then A^{-} clearly coincides with the ordinary inverse $A^{-1} : AD \to D$.

The properties of a quasi-inverse are similar to those of an ordinary inverse; indeed, we have the following proposition:

Lemma 1.1.1

(i) If $x \in D$, then $z \in Ax$, iff $x \in A^{-}z$.

(ii) $(A^{-}(AD)^{\circ})^{\circ} = D$.

(iii) For the set mapping $(A^{-})^{-} : D \to \mathfrak{S}(Y)$ we have $(A^{-})^{-} = A$.

Proof: (i) is an immediate consequence of the definition (1.1.2). To prove (ii), observe that, for any $y \in (AD)^{\circ}$, $A^{-}y \subset D$; hence, $(A^{-}(AD)^{\circ})^{\circ} = \bigcup_{z \in (AD)^{\circ}} A^{-}z \subset D$. Conversely, let $x \in D$. Choosing arbitrarily $y \in Ax \subset (AD)^{\circ}$, we have by (i), $x \in A^{-}y \subset (A^{-}(AD)^{\circ})^{\circ}$; thus, $D \subset (A^{-}(AD)^{\circ})^{\circ}$. As for (iii), choose a fixed $x \in D$. If $y \in Ax$, then $y \in (AD)^{\circ}$ and $x \in \{u : u \in D, y \in Au\} = A^{-}y$. Then, however, $y \in \{z : z \in (AD)^{\circ}, x \in A^{-}z\} = (A^{-})^{-}x$. Hence, $Ax \subset (A^{-})^{-}x$. Conversely, let $y \in (A^{-})^{-}x = \{z : z \in (AD)^{\circ}, x \in A^{-}z\}$; then $y \in (AD)^{\circ}$ and $x \in A^{-}y = \{u : u \in D, y \in Au\}$. Thus, $x \in D$ and $y \in Ax$, i.e., $(A^{-})^{-}x \subset Ax$ which concludes the proof.

Definition: Let $A : X \to \mathfrak{S}(Y)$ be a set mapping, and let $D \subset X$, $D \neq \emptyset$; A will be called simple on D, if

$$x_1, x_2 \in D,\ x_1 \neq x_2 \Rightarrow (Ax_1) \cap (Ax_2) = \emptyset. \tag{1.1.3}$$

Lemma 1.1.2. A is simple on $D \Leftrightarrow$ the quasi-inverse $A^{-} : (AD)^{\circ} \to \mathfrak{S}(D)$ is an operator on $(AD)^{\circ}$.

Proof: a) Let A be simple on D. Choose arbitrarily $y \in (AD)^{\circ}$ and assume that $x_1, x_2 \in A^{-}y$. Then we have by (i) in Lemma 1.1.1, $y \in Ax_1$ and $y \in Ax_2$, so that $(Ax_1) \cap (Ax_2) \neq \emptyset$; hence, by (1.1.3), $x_1 = x_2$, i.e., A^{-} is an operator on $(AD)^{\circ}$.

b) Conversely, let A^{-} be an operator on $(AD)^{\circ}$. Suppose that for some $x_1, x_2 \in D$ we have $(Ax_1) \cap (Ax_2) \neq \emptyset$. Then there exists $y \in (AD)^{\circ}$ such that $y \in Ax_1$ and $y \in Ax_2$, i.e., by (i) in Lemma 1.1.1, $x_1 \in A^{-}y$, $x_2 \in A^{-}y$. However, since $A^{-}y$ is a singleton, it follows that $x_1 = x_2$; hence, (1.1.3) holds, i.e., A is simple on D.

Examples of simple mappings $A, B : R^1 \to \mathfrak{S}(R^1)$ are given in Figure 1.1.

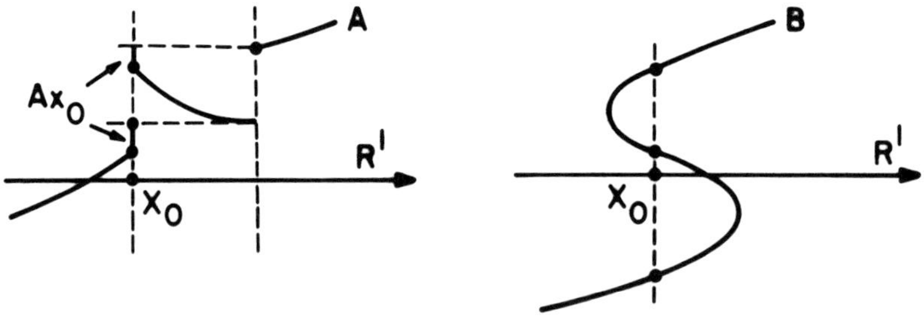

Figure 1.1

Next, if $A : X \to \mathfrak{S}(Y)$ is a set mapping and $B : Y \to Z$ is an operator, we define the set mapping $BA : X \to \mathfrak{S}(Z)$ by $(BA)x = B(Ax) \subset Z$ for each $x \in X$.

Then, for any $D \subset X$, $D \neq \emptyset$, we have

$$((BA)D)^{\circ} = B(AD)^{\circ}, \tag{1.1.4}$$

because $\bigcup_{x \in D} (BA)x = \bigcup_{x \in D} B(Ax) = B(\bigcup_{x \in D} Ax)$.

Moreover, if $C : U \to X$ is an operator, the set mapping AC is defined in a similar way, and, for any $D' \subset U$, $D' \neq \emptyset$, we have

$$((AC)D')^{\circ} = (A(CD'))^{\circ}. \tag{1.1.5}$$

Lemma 1.1.3: Let X,Y,Z,U be nonempty sets, and let $A : X \to \mathfrak{S}(Y)$.

(a) If $B : Y \to Z$ is 1-to-1, then BA is simple on $X \Leftrightarrow A$ is simple on X.

(b) If $C : U \to X$ is 1-to-1 and onto, then AC is simple on $U \Leftrightarrow A$ is simple on X.

Proof: (a) Let A be simple on X, and assume that $((BA)x_1) \cap ((BA)x_2) \neq \emptyset$ for some $x_1,x_2 \in X$; then there exists $y_1 \in Ax_1$ and $y_2 \in Ax_2$ such that $By_1 = By_2$. Thus, $y_1 = y_2$, i.e., $(Ax_1) \cap (Ax_2) \neq \emptyset$; consequently, $x_1 = x_2$, i.e., BA is simple on X.

Conversely, let BA be simple on X, and assume that $(Ax_1) \cap (Ax_2) \neq \emptyset$ for some $x_1,x_2 \in X$; then there exists y such that $y \in Ax_1$, $y \in Ax_2$ so that $By \in (BA)x_1$ and $By \in (BA)x_2$. Hence, $((BA)x_1) \cap ((BA)x_2) \neq \emptyset \Rightarrow x_1 = x_2$; thus A is simple on X.

(b) Let A be simple on X, and let $u_1,u_2 \in U$, $u_1 \neq u_2$; then $Cu_1 \neq Cu_2$, and consequently, the sets $(AC)u_1$ and $(AC)u_2$ have an empty intersection. Hence, AC is simple on U.

Conversely, let AC be simple on U, and assume that $(Ax_1) \cap (Ax_2) \neq \emptyset$ for some $x_1,x_2 \in X$; since C is 1-to-1 from U onto X, there exist unique elements $u_1,u_2 \in U$ such that $x_1 = Cu_1$ and $x_2 = Cu_2$. Thus, we have $((AC)u_1) \cap ((AC)u_2) \neq \emptyset$, which implies that $u_1 = u_2$; consequently $x_1 = x_2$, i.e., A is simple on X.

If $A,B : X \to \mathfrak{S}(Y)$ are set mappings and Y is a linear space over the field of real or complex numbers, we define the set mappings αA, $(A+B) : X \to \mathfrak{S}(Y)$ (with α being a scalar) by

$$(\alpha A)x = \alpha(Ax) = \{\alpha z : z \in Ax\}. \tag{1.1.6}$$

$$(A+B)x = Ax + Bx = \{y+z : y \in Ax, z \in Bx\}. \qquad (1.1.7)$$

Let now $\mathcal{L}$ be a fixed nontrivial linear space over the field of real or complex numbers, and let M,N be linear subspaces of $\mathcal{L}$ such that $\mathcal{L} = M \oplus N$. (As known, $\mathcal{L} = M\oplus N \Leftrightarrow \mathcal{L} = M+N$ and $M \cap N = \{0\}$.)

Furthermore, let P be the projection from $\mathcal{L}$ onto N along M. Note that P is a linear operator having the properties 1 : (1) $P\mathcal{L} = N$, (2) $P^2 = P$, (3) $PM = \{0\}$. Then we have the following asserties:

Lemma 1.1.4: Let M,N be linear subspaces of $\mathcal{L}$ with $\mathcal{L} = M\oplus N$, let S be a non-empty subset of $\mathcal{L}$ and let P be the projection of $\mathcal{L}$ onto N along M. Then

$$M+S = P^{-1}\{PS\}. \qquad (1.1.8)$$

Proof: Let $x \in M+S$; then there exists $y \in M$ and $z \in S$ such that $x = y+z$. Since $Py = 0$, we have $Px = Pz \in PS$; hence; $x \in P^{-1}\{PS\}$. Conversely, let $x \in P^{-1}\{PS\}$; then $Px \in PS$, i.e., there exists $z \in S$ such that $Px = Pz$. Thus, $P(x-z) = 0 \Rightarrow x-z = y \in M$. Hence, $x = y+z \in M+S$.

By a Hilbert space $\mathcal{H}$ we will mean a real or complex (not necessarily separable) complete inner product space.

Let $\tilde{\mathcal{H}}$ be a real Hilbert space and let $M : \tilde{\mathcal{H}} \to \mathfrak{S}(\tilde{\mathcal{H}})$ be a set mapping; M is called monotone on $\tilde{\mathcal{H}}$, if for all $x_1,x_2 \in \tilde{\mathcal{H}}$ and all $z_1 \in Mx_1$, $z_2 \in Mx_2$ we have

$$\langle z_1-z_2, x_1-x_2\rangle \geq 0. \qquad (1.1.9)$$

Moreover, M is called maximal monotone on $\tilde{\mathcal{H}}$, if $M' : \tilde{\mathcal{H}} \to \mathfrak{S}(\tilde{\mathcal{H}})$ monotone, $M'x \supset Mx$ for all $x \in \tilde{\mathcal{H}}$ implies that $M' = M$.

Let $M : \tilde{\mathcal{H}} \to \mathfrak{S}(\tilde{\mathcal{H}})$ be a set mapping ; we say that M is coercive, if

$$\lim_{a\to\infty} a^{-1}\inf\{\langle z,x\rangle : z \in Mx,\ x \in \tilde{\mathcal{H}},\ \|x\| \geq a\} = \infty. \qquad (1.1.10)$$

The concepts of maximal monotonicity and coerciveness can be combined to yield the following powerful result due to Rockafellar:

Lemma 1.1.5. If $\tilde{\mathcal{H}}$ is a real Hilbert space and $M : \tilde{\mathcal{H}} \to \mathfrak{S}(\tilde{\mathcal{H}})$ is maximal monotone and coercive on $\tilde{\mathcal{H}}$, then $(M\tilde{\mathcal{H}})^{\circ} = \tilde{\mathcal{H}}$. (For the proof see [2]).

If M is a monotone operator and M is continuous, then M is automatically maximal monotone. Indeed, we have the proposition:

Lemma 1.1.6: Let $\mathfrak{H}$ be a real Hilbert space, and let $M : \mathfrak{H} \to \mathfrak{H}$ be a monotone operator; if M is continuous, then it is maximal monotone on $\mathfrak{H}$.

Proof: Let $M' : \mathfrak{H} \to \mathfrak{S}(\mathfrak{H})$ be a set mapping such that $M'x \supset \{Mx\}$ for every $x \in \mathfrak{H}$, and assume that, for at least one point $x_o \in \mathfrak{H}$, $M'x_o \neq \{Mx_o\}$. We are going to show that M' is not monotone. Indeed, choose some $z \in M'x_o$, $z \neq Mx_o$, and denote $m = \|z-Mx_o\| > 0$. If $\xi \in \mathfrak{H}$ is an arbitrary point, we have

$$\langle z-M\xi,x_o-\xi\rangle = \langle z-Mx_o,x_o-\xi\rangle + \langle Mx_o-M\xi,x_o-\xi\rangle. \qquad (1.1.11)$$

Now, M is continuous at x_o, i.e., there exists $\delta > 0$ such that

$$\|Mx_o-M\xi\| < \frac{m}{2}$$

whenever $\|x_o-\xi\| \leq \delta$. Define a point ξ_o by

$$\xi_o = x_o + \frac{\delta}{m}(z-Mx_o).$$

Then we have $x_o-\xi_o = \frac{\delta}{m}(z-Mx_o)$, so that $\|x_o-\xi_o\| = \frac{\delta}{m}\|z-Mx_o\| = \delta$; consequently $\|Mx_o-M\xi_o\| < \frac{m}{2}$. Thus, by monotonicity of M and Schwartz inequality,

$$0 \leq \langle Mx_o-M\xi_o,x_o-\xi_o\rangle < \frac{m}{2}\delta.$$

On the other hand,

$$\langle z-Mx_o,x_o-\xi_o\rangle = -\frac{\delta}{m}\|z-Mx_o\|^2 = -\delta m;$$

hence, by (1.1.11),

$$\langle z-M\xi_o,x_o-\xi_o\rangle < -\delta m + \tfrac{1}{2}\delta m = -\tfrac{1}{2}\delta m < 0. \qquad (1.1.12)$$

Since $M'\xi_o \supset \{M\xi_o\}$, inequality (1.1.12) shows that M' is not monotone. Hence, there does not exist any proper monotone extension of M, i.e., M is maximal.

Finally, we will use the following notation: If $\mathfrak{L}$ and $\mathfrak{L}'$ are normed linear spaces, then $[\mathfrak{L},\mathfrak{L}']$ will signify the space of all linear bounded operators from $\mathfrak{L}$ into $\mathfrak{L}'$. Moreover, if $\alpha > 0$, then $\mathrm{Lip}_\alpha[\mathfrak{L},\mathfrak{L}']$ will denote the set of all (not necessarily linear) operators $A : \mathfrak{L} \to \mathfrak{L}'$ such that

$$\|Ax_1 - Ax_2\|' \leq \alpha \|x_1 - x_2\| \tag{1.1.13}$$

for all $x_1, x_2 \in \mathcal{L}$.

Problems

1.1.1 It is meaningless to speak about a "linear set mapping A" unless A is an operator, i.e., prove: Let X,Y be linear spaces, and let $A : X \to \mathfrak{S}(Y)$ satisfy the conditions

(i) $A(\lambda x) = \lambda(Ax)$ for each $x \in X$ and each scalar λ,

(ii) $A(x+y) = Ax+Ay$ for all $x,y \in X$.

Then A is an operator.

1.1.2 If $\mathcal{L}$ is a linear space and $x,x' \in \mathcal{L}$ are distinct, we let $[x,x'] = \{\lambda x+(1-\lambda)x' : \lambda \in [0,1]\}$ (a segment). A subset $D \subset \mathcal{L}$ will be called connected, if

(i) D contains at least two points.

(ii) for any distinct points, $x,x' \in D$ there exist points $x_1,x_2,\ldots,x_{n-1} \in D$ such that $[x_k,x_{k+1}] \subset D$ for $k = 0,1,\ldots,n-1$, where $x_o = x$ and $x_n = x'$.

Furthermore, let $\mathcal{L},\mathcal{L}'$ be normed linear spaces, let $D \subset \mathcal{L}$ be nonempty and let $A : D \to \mathcal{L}'$; the operator A is said to satisfy the Hölder condition with exponent $p > 0$, if, for some $\alpha > 0$ and all $x_1,x_2 \in D$, we have

(*) $\qquad \|Ax_1 - Ax_2\|' \leq \alpha \|x_1 - x_2\|^p.$

Clearly, if $p < 1$, this condition is weaker in a certain sense than (1.1.13). On the other hand, (*) is "practically meaningful" only if $p \leq 1$, i.e., we have the proposition: If D is connected and A satisfies (*) with $p > 1$, then A is a "constant mapping", i.e., there exist $y \in \mathcal{L}'$ such that $AD = \{y\}$.

1.1.3 Let $T \geq 0$, let c be real and let $K = \{x : x$ absolutely continuous on $[0,T]$, $x(0) = c$, $x' \in L_2[0,T]\}$, where $L_2[0,T]$ is assumed to be real. Let the operator $A : K \to L_2[0,T]$ be defined by $(Ax)(t) = x'(t)$; then A is monotone on K.

1.2 Abstract Network, Solution

In this section, we introduce the concept of an abstract network and give theorems on the existence and uniqueness of a solution. We then discuss a special case - the abstract Hilbert network.

Let $\mathcal{L}$ be a fixed nontrivial linear space over the field of real or complex numbers, and let M, N be linear subspaces of $\mathcal{L}$ such that $\mathcal{L} = M \oplus N$. Moreover, let D be a nonempty subset of $\mathcal{L}$, and let $Z : D \to \mathfrak{S}(\mathcal{L})$ be a set mapping. Then the ordered triplet $\mathfrak{n} = [Z,N,M]$ will be called an abstract network (a.n.) over $\mathcal{L}$.

Definition. Let $\mathfrak{n} = [Z,N,M]$ be an a.n. over $\mathcal{L}$, and let $e \in \mathcal{L}$; an element $i \in \mathcal{L}$ will be called a solution of $\mathfrak{n}$ corresponding to e, if

(i) there exists $v \in Zi$ such that

$$v-e \in M \tag{1.2.1}$$

(ii) $i \in N \cap D$.

Let $\tilde{D} \subset D$; if $N \cap \tilde{D} \neq \emptyset$, define the set $Q(\tilde{D})$ by

$$Q(\tilde{D}) = M+(Z(N\cap\tilde{D}))^{o}, \tag{1.2.2}$$

and if $N \cap \tilde{D} = \emptyset$, we put $Q(\tilde{D}) = \emptyset$.

Using Lemma 1.1.4 and (1.1.4) it follows that

$$Q(\tilde{D}) = P^{-1}\{[(PZ)(N\cap\tilde{D})]^{o}\}, \tag{1.2.3}$$

where P is the projection from $\mathcal{L}$ onto N along M.

Now we can state the theorem on the existence of a solution.

Theorem 1.2.1. Let $\mathfrak{n} = [Z,N,M]$ be an a.n. over $\mathcal{L}$, let $\tilde{D} \subset D$ and let $e \in \mathcal{L}$; then $\mathfrak{n}$ possesses a solution i corrresponding to e and $i \in \tilde{D}$, iff $e \in Q(\tilde{D})$. In this case, the set I of all solutions in $\tilde{D}$ is given by

$$I = (PZ)^{-}Pe, \tag{1.2.4}$$

where $(PZ)^{-} : [(PZ)(N\cap\tilde{D})]^{o} \to \mathfrak{S}(N\cap\tilde{D})$ is the quasi-inverse of the set mapping $PZ : (N\cap\tilde{D}) \to \mathfrak{S}([(PZ)(N\cap\tilde{D})]^{o})$ and P is the projection from $\mathcal{L}$ onto N along M.

Proof: If $N\cap\tilde{D} = \emptyset$, then $Q(\tilde{D}) = \emptyset$; in this case there exists no solutions in $\tilde{D}$ whatever e can be, and our assertion holds. Thus, assume that $N\cap\tilde{D} \neq \emptyset$.

Let $e \in Q(\tilde{D})$; then, by (1.2.2), there exists $x \in M$ and $y \in (Z(N\cap\tilde{D}))^{o}$ such that $e = x+y$. Consequently, there exists $i \in N\cap\tilde{D}$ such that $y \in Zi$;

thus, $y-e = -x \in M$, i.e., i is a solution of $\mathfrak{n}$ corresponding to e by definition (i), (ii). (Note that $i \in N\cap\tilde{D} \subset N\cap D$).

Conversely, let $i \in \tilde{D}$ be a solution of $\mathfrak{n}$ corresponding to e. Then by (ii), $i \in N\cap\tilde{D}$, and by (i), there exists $v \in Zi$ such that $v-e \in M$, i.e., $e-v \in M$. Thus $e \in v+M \subset (Z(N\cap\tilde{D}))^{\circ} + M = Q(\tilde{D})$.

In order to prove the relation (1.2.4), let I be the set of all solutions in $\tilde{D}$ which correspond to e. Also, note that the set $(PZ)^{-}Pe$ is well-defined. Indeed, since $e \in Q(\tilde{D})$, it follows by (1.2.3) that $Pe \in [(PZ)(N\cap\tilde{D})]^{\circ}$, i.e., Pe is in the domain of $(PZ)^{-}$.

Let $i \in (PZ)^{-}Pe$. Since $(PZ)^{-}$ maps its domain into $\mathfrak{S}(N\cap\tilde{D})$, we have $i \in N\cap\tilde{D} \subset N\cap D$, so that i satisfies (ii). Moreover, by (i) in Lemma 1.1.1, $Pe \in PZi$; thus there exists $v \in Zi$ such that $Pe = Pv$. Consequently, $v-e \in M$, i.e., i satisfies (i). Hence, $i \in I$ which implies that $(PZ)^{-}Pe \subset I$.

Conversely, let $i \in I$. Then, by (ii), $i \in N\cap\tilde{D}$, and by (i), there exists $v \in Zi$ such that $v-e \in M$. Hence $Pe = Pv$, i.e., $Pe \in PZi$. Thus, since i is in the domain of PZ, we have by (i) in Lemma 1.1.1, $i \in (PZ)^{-}Pe$. Consequently, $I \subset (PZ)^{-}Pe$, which proves (1.2.4).

From the physical point of view, the most important situation occurs when every solution in $\tilde{D}$ is determined uniquely. Thus, let us define the following concept:

Let $\tilde{D} \subset D$ and assume that $N\cap\tilde{D} \neq \emptyset$; the abstract network $\mathfrak{n}$ will be called regular on $\tilde{D}$, if for each $e \in Q(\tilde{D})$ the network $\mathfrak{n}$ possesses in $\tilde{D}$ a unique solution i corresponding to e.

Then we have the following assertion:

Theorem 1.2.2. Let $\mathfrak{n} = [Z,N,M]$ be an a.n. over $\mathfrak{L}$, let P be the projection from $\mathfrak{L}$ onto N along M, and let $\tilde{D} \subset D$ be such that $N\cap\tilde{D} \neq \emptyset$; then $\mathfrak{n}$ is regular on $\tilde{D}$, iff the set mapping PZ is simple on $N\cap\tilde{D}$.

Proof: Using our definition and Lemma 1.1.2, we have: $\mathfrak{n}$ is regular on $\tilde{D} \Leftrightarrow$ for each $e \in Q(\tilde{D})$ the solution $i \in \tilde{D}$ is unique $\Leftrightarrow$ for each $e \in Q(\tilde{D})$ the set I is a singleton $\Leftrightarrow (PZ)^{-}$ is an operator on $[(PZ)(N\cap\tilde{D})]^{\circ} \Leftrightarrow$ PZ is simple on $N\cap\tilde{D}$.

Consider now the following situation: Suppose that i is a solution of $\mathfrak{n}$ corresponding to some $e \in Q(D)$, where Q(D) is defined by (1.2.2) with $\tilde{D}$ replaced by D. In general, $\mathfrak{n}$ need not be regular on the entire set D; therefore, we may look for a subdomain $\tilde{D}_i$ of D, as large as possible, such that $\tilde{D}_i$ contains i and $\mathfrak{n}$ is regular on $\tilde{D}_i$. The existence of such a subdomain $\tilde{D}_i$ is guaranteed by the following result:

Theorem 1.2.3. Let $e \in Q(D)$ and let $i \in D$ be a fixed solution of $\mathfrak{n}$ corresponding to e; then there exists $\tilde{D}_i \subset D$ such that

(i) $i \in \tilde{D}_i$,

(ii) $\mathfrak{n}$ is regular on $\tilde{D}_i$,

(iii) $\tilde{D}_i$ is maximal, i.e., $\mathfrak{n}$ is not regular on any other $\tilde{D}$ which properly contains $\tilde{D}_i$.

Proof: Let $S_i = \{\tilde{D}^\alpha : \tilde{D}^\alpha \subset D,\ i \in \tilde{D}^\alpha,$ PZ is simple on $N \cap \tilde{D}^\alpha\}$. The collection S_i is nonempty, since $\tilde{D}^o = \{i\} \in S_i$. Moreover, S_i is partially ordered by the set inclusion; also, if $\mathfrak{J} = \{\tilde{D}^\alpha : \alpha \in I\} \subset S_i$ is a chain, then clearly the set $\bigcup_{\alpha \in I} \tilde{D}^\alpha \in S_i$ and is an upper bound for $\mathfrak{J}$. Hence, by Zorn's Lemma [1], there exists a maximal element $\tilde{D}_i$ in S_i; then Theorem 1.2.2 concludes the proof.

Remark 1. Even in the case that $i \in D$ is the unique solution of $\mathfrak{n}$ corresponding to some $e \in Q(D)$, the maximal subdomain $\tilde{D}_i$ need not be determined uniquely. This fact is illustrated by the example $N \cap D = R^1$ and $(PZ)x = x^2$.

Reviewing Theorems 1.2.1-1.2.3 we see that the assumption $\tilde{D} = D$ entails no loss of generality, because, if necessary, we can take for D some subset of the domain of Z in order to have regularity. Thus, in the sequel we will deal only with D. Moreover, the set mapping $A : Q(D) \to \mathfrak{S}(D)$, defined by $A = (PZ)^- P$ (see Theorem 1.2.1), will be called the admittance of $\mathfrak{n}$. Note that, due to Theorem 1.2.2, A is an operator on $Q(D) \Leftrightarrow \mathfrak{n}$ is regular on D.

Particularly important is the case when $Q(D) = \mathfrak{L}$, i.e., for any $e \in \mathfrak{L}$ the network $\mathfrak{n}$ possesses a solution. Here, we have the proposition:

Theorem 1.2.4. For an a.n. $\mathfrak{n}$,

$$Q(D) = \mathcal{L} \Leftrightarrow [(PZ)(N \cap D)]^{\circ} = N.$$

Proof: If $[\ldots]^{\circ} = N$, then clearly $Q(D) = P^{-1}\{[\ldots]^{\circ}\} = \mathcal{L}$. Conversely, if $Q(D) = P^{-1}\{[\ldots]^{\circ}\} = \mathcal{L}$, then necessarily $[(PZ)(N \cap D)]^{\circ} \supset N$; however, since $[\ldots]^{\circ} = P[Z(N \cap D)]^{\circ} \subset N$, we have equality and our claim is proved.

Let us now consider a particular case of an abstract network, i.e., the abstract Hilbert network.

Let $\mathcal{H}$ and $\mathcal{H}'$ be fixed Hilbert spaces, and let $a \in [\mathcal{H},\mathcal{H}']$ be a given operator. Denote

$$N_a = \{x : x \in \mathcal{H},\ ax = 0\}.$$

Clearly, N_a is a closed linear subspace of $\mathcal{H}$; also, note that $\mathcal{H} = N_a^{\perp} \oplus N_a$.

Now, the abstract network $\mathfrak{n} = [Z, N_a, N_a^{\perp}]$ over $\mathcal{H}$ will be called an abstract Hilbert network over $\mathcal{H}$.

Since the subspace N_a and $N_a^{\perp}$ are completely determined by the operator a, we will denote the abstract Hilbert network by the ordered pair (Z,a).

Also, note the fact that, due to the relation $\mathcal{H} = N_a^{\perp} \oplus N_a$, the projection P from $\mathcal{H}$ onto N_a along $N_a^{\perp}$ is the orthogonal projection from $\mathcal{H}$ onto N_a, and has the properties (1) $P\mathcal{H} = N_a$, (2) $P^2 = P$, (3) $PN_a^{\perp} = \{0\}$, (4) $P^* = P$ (P^* being the adjoint of P), (5) $\|P\| = 1$.

Recalling the definition of a solution of an abstract network, we see readily that a solution i of an abstract Hilbert network $\mathfrak{n} = (Z,a)$ over $\mathcal{H}$ corresponding to $e \in \mathcal{H}$ can be defined by the relations:

K_1^* : there exists $v \in Zi$ such that $v - e \in N_a^{\perp}$,

K_2^* : $i \in N_a \cap D$.

Alternatively, if $\mathfrak{n} = (Z,a)$ is an a.H.n. over $\mathcal{H}$ and $e \in \mathcal{H}$, then an element $i \in \mathcal{H}$ is a solution of $\mathfrak{n}$ corresponding to e, if $i \in D$ and

(i)* there exists $v \in Zi$ such that $\langle c, v-e \rangle = 0$ for all $c \in \mathcal{H}$ with $ac = 0$,

(ii)* $ai = 0$.

It is clear that conditions (i)* and (ii)* are straightforward abstractions of Kirchhoff's loop and node law, respectively.

Note also the following fact: the concept of an abstract network is purely algebraic, and theorems 1.2.1-1.2.3 use only the set-theoretic properties of the mappings involved. On the other hand, when dealing with an abstract Hilbert network, we have the structure of a Hilbert space at our disposal, and consequently, we will be able to prove theorems on existence and uniqueness of a solution taking full advantage of this fact. Thus, we now prove some theorems concerning abstract Hilbert networks which use the topological properties of the mappings Z and a.

As before, we will assume that $\widetilde{D} = D$, and that $N_a \cap D \neq \emptyset$.

Theorem 1.2.5. Let $\mathfrak{n} = (Z,a)$ be an a.H.n. over $\mathcal{H}$.

(a) If for all $x_1, x_2 \in N_a \cap D$, $x_1 \neq x_2$ and all $y_1 \in Zx_1$, $y_2 \in Zx_2$ we have

$$\langle y_1 - y_2, x_1 - x_2 \rangle \neq 0, \tag{1.2.5}$$

then $\mathfrak{n}$ is regular on D. If, in addition, Z is an operator, then

$$\langle Ae_1 - Ae_2, e_1 - e_2 \rangle \neq 0 \tag{1.2.6}$$

for all $e_1, e_2 \in Q(D)$ such that $Pe_1 \neq Pe_2$, where A is the admittance of $\mathfrak{n}$.

(b) If there exist constants $c > 0$ and $p > 1$ such that for any $x_1, x_2 \in N_a \cap D$ and any $y_1 \in Zx_1$, $y_2 \in Zx_2$ we have

$$|\langle y_1 - y_2, x_1 - x_2 \rangle| \geq c\|x_1 - x_2\|^p, \tag{1.2.7}$$

then $\mathfrak{n}$ is regular on D; moreover,

$$\|Ae_1 - Ae_2\| \leq c^{-\frac{1}{p-1}} \|P(e_1 - e_2)\|^{\frac{1}{p-1}} \tag{1.2.8}$$

and

$$|\langle Ae_1 - Ae_2, e_1 - e_2 \rangle| \leq c^{-\frac{1}{p-1}} \|P(e_1 - e_2)\|^{\frac{p}{p-1}} \tag{1.2.9}$$

for all $e_1, e_2 \in Q(D)$.

(c) If there exist constants $c > 0$ and $p > 1$ such that for any $x_1, x_2 \in N_a \cap D$ and any $y_1 \in Zx_1$, $y_2 \in Zx_2$ we have

$$\mathrm{Re}\langle y_1 - y_2, x_1 - x_2 \rangle \geq c\|x_1 - x_2\|^p, \tag{1.2.10}$$

then $\mathfrak{h}$ is regular on D, (1.2.8), (1.2.9) hold and

$$\mathrm{Re}\langle Ae_1-Ae_2,e_1-e_2\rangle \geq 0 \tag{1.2.11}$$

for all $e_1,e_2 \in Q(D)$.

Proof: (a) Since $Px_j = x_j$ for $x_j \in N_a \cap D$, $j = 1,2$, and $P^* = P$, (1.2.5) yields

$$\langle Py_1-Py_2,x_1-x_2\rangle \neq 0 \tag{1.2.12}$$

whenever $x_1 \neq x_2$ and $y_j \in Zx_j$. Let $x_1,x_2 \in N_a \cap D$, $x_1 \neq x_2$ and suppose that $(PZx_1)\cap(PZx_2) \neq \emptyset$. Then there exist elements $y_1 \in Zx_1$ and $y_2 \in Zx_2$ such that $Py_1 = Py_2$. This, however, contradicts (1.2.12); hence, necessarily $(PZx_1)\cap(PZx_2) = \emptyset$, i.e., PZ is simple on $N_a \cap D$, and consequently, by Theorem 1.2.2, $\mathfrak{h}$ is regular on D.

Next assume that Z is an operator; then (1.2.5) means that

$$\langle Zx_1-Zx_2,x_1-x_2\rangle \neq 0 \tag{1.2.13}$$

whenever $x_1,x_2 \in N_a \cap D$, $x_1 \neq x_2$. Thus, we have as before for such x_1,x_2,

$$\langle PZx_1-PZx_2,x_1-x_2\rangle \neq 0. \tag{1.2.14}$$

Also, $PZ : N_a \cap D \to (PZ)(N_a \cap D)$ is a 1-to-1 onto operator. Choose $y_1,y_2 \in (PZ)(N_a \cap D)$, $y_1 \neq y_2$ and put $x_1 = (PZ)^{-1}y_1$, $x_2 = (PZ)^{-1}y_2$. Since $x_1 \neq x_2$, we have by (1.2.14),

$$\langle y_1-y_2,(PZ)^{-1}y_1 - (PZ)^{-1}y_2\rangle \neq 0. \tag{1.2.15}$$

Now, let $e_j \in Q(D)$, $j = 1,2$ be such that $Pe_1 \neq Pe_2$; since $Pe_j \in (PZ)(N_a \cap D)$ by (1.2.3), we can set $y_j = Pe_j$ into (1.2.15) and get

$$\begin{aligned} 0 &\neq \langle Pe_1-Pe_2,(PZ)^{-1}Py_1 - (PZ)^{-1}Py_2\rangle \\ &= \langle Pe_1-Pe_2,Ae_1-Ae_2\rangle. \end{aligned} \tag{1.2.16}$$

Noting the fact that $PAe = Ae$ for any $e \in Q(D)$, (since $Ae \in N_a$), we conclude from (1.2.16) readily that (1.2.6) holds.

(b) Since (1.2.7) implies (1.2.5), $\mathfrak{h}$ is regular on D by (a). If $x_j \in N_a \cap D$ and $y_j \in Zx_j$, $j = 1,2$, then $Px_j = x_j$ and we obtain from (1.2.7) by Schwartz inequality,

$$\begin{aligned} c\|x_1-x_2\|^p &\leq |\langle y_1-y_2,P(x_1-x_2)\rangle| = |\langle Py_1-Py_2,x_1-x_2\rangle| \\ &\leq \|Py_1-Py_2\|\cdot\|x_1-x_2\|. \end{aligned}$$

Hence,

$$\|x_1-x_2\| \leq c^{-\frac{1}{p-1}}\|Py_1-Py_2\|^{\frac{1}{p-1}} \tag{1.2.17}$$

for $x_j \in N_a \cap D$, $y_j \in Zx_j$, $j = 1,2$.

Now, choose $e_j \in Q(D)$, $j = 1,2$ and put $i_j = Ae_j = (PZ)^- Pe_j \in N_a \cap D$. By definition of $(PZ)^-$, $Pe_j \in (PZ)i_j$; hence, there exists $y_j \in Zi_j$ such that $Py_j = Pe_j$ for $j = 1,2$. Putting $x_j = i_j$ into (1.2.17), we obtain

$$\|Ae_1-Ae_2\| \leq c^{-\frac{1}{p-1}}\|Pe_1-Pe_2\|^{\frac{1}{p-1}}$$

and (1.2.8) is proved.

To prove (1.2.9), note that $PAe = Ae$ for any $e \in Q(D)$; thus, for any $e_1,e_2 \in Q(D)$,

$$|\langle Ae_1-Ae_2,e_1-e_2\rangle| = |\langle Ae_1-Ae_2,Pe_1-Pe_2\rangle| \leq \|Ae_1-Ae_2\|\cdot\|Pe_1-Pe_2\|.$$

This and (1.2.8) finishes the proof.

(c) Since (1.2.10) implies (1.2.7), $\mathfrak{n}$ is regular and (1.2.8), (1.2.9) hold by proposition (b). As before, (1.2.10) implies that

$$\mathrm{Re}\langle Py_1-Py_2,x_1-x_2\rangle \geq 0 \tag{1.2.18}$$

for $x_j \in N_a \cap D$, $y_j \in Zx_j$, $j = 1,2$. Choose $e_j \in Q(D)$, $j = 1,2$ and put $i_j = Ae_j = (PZ)^- Pe_j \in N_a \cap D$. Then $Pe_j \in (PZ)i_j$, i.e., there exists $y_j \in Zi_j$ such that $Pe_j = Py_j$. Putting $x_j = i_j$ into (1.2.18) we get
$0 \leq \mathrm{Re}\langle Pe_1-Pe_2,Ae_1-Ae_2\rangle = \mathrm{Re}\langle e_1-e_2,Ae_1-Ae_2\rangle$, which concludes the proof.

Remark 2. The inequality (1.2.6) need not hold, if Z is not an operator. This fact can easily be demonstrated by considering the set mapping $PZ : R^1 \to \mathfrak{S}(R^1)$ whose graph is depicted in Figure 1.2. We understand that $N_a \cap D = R^1$. Then the set mapping Z clearly satisfies the condition (1.2.5), (P is an identity on N_a!), but for points $y_1,y_2 \in R^1$ we have $\langle (PZ)^- y_1 - (PZ)^- y_2,y_1-y_2\rangle = 0$. This implies that, for certain $e_1,e_2 \in Q(D)$ with $Pe_1 \neq Pe_2$, $\langle Ae_1-Ae_2,e_1-e_2\rangle = 0$.

Remark 3. A comment on the assumption (1.2.10) made in Theorem 1.2.5 is in order. Most specific networks have the property that D is a set containing line segments. In this case, a set mapping $Z : D \to \mathfrak{S}(\mathcal{H})$ can satisfy the

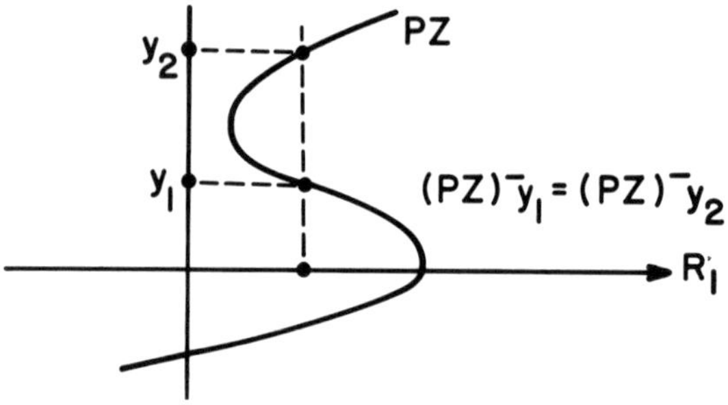

Figure 1.2

condition (1.2.10) only if $p \geq 2$. In order to see this, consider the following. If $x_1, x_2 \in \mathcal{H}$, $x_1 \neq x_2$, then the set $[x_1,x_2] = \{\lambda x_1 + (1-\lambda)x_2 : \lambda \in [0,1]\}$ will be called a segment. We now have the

Proposition: Let $D \subset \mathcal{H}$ be a subset which contains a segment, and let $c > 0$, $1 < p < 2$; then there does not exist a set mapping $A : D \to \mathfrak{S}(\mathcal{H})$ which satisfies the inequality

$$\operatorname{Re}\langle y_1 - y_2, x_1 - x_2 \rangle \geq c\|x_1 - x_2\|^p \tag{1.2.19}$$

for all $x_1, x_2 \in D$ and $y_1 \in Ax_1$, $y_2 \in Ax_2$.

Proof: Suppose that such a mapping exists. Let $[x,x'] \subset D$, $x \neq x'$ be a segment existing by our assumption. Choose arbitrarily points $y \in Ax$ and $y' \in Ax'$, and keep them fixed throughout the following consideration.

Next, for each $x_\alpha \in D$, $x_\alpha \neq x, x'$ choose a point $y_\alpha \in Ax_\alpha$, and define an operator $M : D \to \mathcal{H}$ by

$$\begin{aligned} &Mx_\alpha = y_\alpha \text{ for } x_\alpha \neq x, x', \\ &Mx = y,\ Mx' = y'. \end{aligned} \tag{1.2.20}$$

Then from (1.2.19) it follows that

$$\operatorname{Re}\langle Mx_1 - Mx_2, x_1 - x_2 \rangle \geq c\|x_1 - x_2\|^p \tag{1.2.21}$$

for all $x_1, x_2 \in D$.

Furthermore, define the function $\varphi : [0,1] \to R^1$ by

$$\varphi(\lambda) = \operatorname{Re}\langle M(\lambda x' + (1-\lambda)x), x' - x \rangle. \tag{1.2.22}$$

Choose $\lambda_1, \lambda_2 \in [0,1]$; then we have by (1.2.22),

$$(\varphi(\lambda_1) - \varphi(\lambda_2))(\lambda_1 - \lambda_2) \tag{1.2.23}$$
$$= \mathrm{Re}\langle M(\lambda_1 x' + (1-\lambda_1)x) - M(\lambda_2 x' + (1-\lambda_2)x), (x'-x)(\lambda_1-\lambda_2)\rangle.$$

However, letting $x_{\lambda_j} = \lambda_j x' + (1-\lambda_j)x$, $j = 1,2$, we have

$$x_{\lambda_1} - x_{\lambda_2} = (\lambda_1 - \lambda_2)(x'-x); \tag{1.2.24}$$

thus, (1.2.23) yields by (1.2.21),

$$(\varphi(\lambda_1) - \varphi(\lambda_2))(\lambda_1 - \lambda_2) = \mathrm{Re}\langle Mx_{\lambda_1} - Mx_{\lambda_2}, x_{\lambda_1} - x_{\lambda_2}\rangle \tag{1.2.25}$$
$$\geq c\|x_{\lambda_1} - x_{\lambda_2}\|^p = c\|x'-x\|^p |\lambda_1 - \lambda_2|^p.$$

Next, if $\lambda_2 > \lambda_1$, inequality (1.2.25) shows that

$$\varphi(\lambda_2) - \varphi(\lambda_1) \geq \gamma(\lambda_2 - \lambda_1)^{p-1} > 0, \tag{1.2.26}$$

where we have put $\gamma = c\|x'-x\|^p > 0$, and where $0 < p-1 < 1$.

Now, choose arbitrarily an integer $n \geq 1$; then by (1.2.26),

$$\varphi(1) - \varphi(0) = (\varphi(1) - \varphi(\tfrac{n-1}{n})) + (\varphi(\tfrac{n-1}{n}) - \varphi(\tfrac{n-2}{n})) + \ldots$$
$$\ldots + (\varphi(\tfrac{1}{n}) - \varphi(0)) \geq n \cdot \gamma(\tfrac{1}{n})^{p-1} = \gamma n^{2-p}.$$

Thus, setting $\lambda_1 = 1$, $\lambda_2 = 0$, we have $x_{\lambda_1} = x$, $x_{\lambda_2} = x'$ and (1.2.25) yields by Schwarz inequality,

$$\|y-y'\| \cdot \|x-x'\| = \|Mx - Mx'\| \cdot \|x-x'\| \geq \mathrm{Re}\langle Mx-Mx', x-x'\rangle \tag{1.2.27}$$
$$= (\varphi(1) - \varphi(0)) \cdot (1-0) \geq \gamma n^{2-p};$$

thence,

$$\|y-y'\| \geq c\|x-x'\|^{p-1} n^{2-p},$$

which is the sought contradiction.

It is interesting to note the following fact: the inequality (1.2.19), (and, obviously, (1.2.7) too), implies that

$$\|y_1 - y_2\| \geq c\|x_1 - x_2\|^{p-1} \tag{1.2.28}$$

for all $x_1, x_2 \in D$ and all $y_1 \in Ax_1$, $y_2 \in Ax_2$. However, in contrast to the above Proposition, it is easy to see that there exist set mappings $A : D \to \mathfrak{S}(\mathcal{H})$, where D contains a segement, such that (1.2.28) is satisfied with $1 < p < 2$.

Indeed, let $\mathcal{H}$ be a nonseparable Hilbert space. Since every orthonormal basis in $\mathcal{H}$ is uncountable, we can choose an uncountable orthonormal system, say $\{e_\lambda : \lambda \in [0,\infty)\}$. Let $D = \{\alpha e_o : \alpha \in [0,1]\}$, and for each $x = \alpha e_o \in D$, let $Ax = e_\alpha$. Clearly, D is a segment, and for any $x_j = \alpha_j e_o \in D$, $j = 1,2$, we have $\|x_1 - x_2\| = \|(\alpha_1 - \alpha_2)e_o\| = |\alpha_1 - \alpha_2| \le 1$; thus for any $p > 1$, we have $\|x_1 - x_2\|^{p-1} \le 1$.

On the other hand, for such x_j,

$$\|Ax_1 - Ax_2\|^2 = \|e_{\alpha_1} - e_{\alpha_2}\|^2 = \langle e_{\alpha_1} - e_{\alpha_2}, e_{\alpha_1} - e_{\alpha_2}\rangle = 2$$

whenever $\alpha_1 \neq \alpha_2$. Hence, we have

$$\|Ax_1 - Ax_2\| \ge \sqrt{2}\, \|x_1 - x_2\|^{p-1}$$

for all $x_1, x_2 \in D$ with $p > 1$, which proves our claim.

Let us now consider more closely the case that $Q(D) = \mathcal{H}$. We can easily prove the following result:

<u>Theorem 1.2.6</u>. Let $\mathcal{H}$ be a real Hilbert space and let $\mathfrak{h} = (Z,a)$ be an a.H.n. over $\mathcal{H}$. Assume that

(i) $D \supset N_a$,

(ii) the set mapping PZ is maximal monotone and coercive on N_a.

Then $Q(D) = \mathcal{H}$, i.e., for any $e \in \mathcal{H}$, $\mathfrak{h}$ possesses a solution corresponding to e.

<u>Proof</u>: First note the fact that N_a, as a closed subspace of $\mathcal{H}$, is a Hilbert space of its own right. Also, due to (i), $N_a \cap D = N_a$. Since PZ maps N_a into $\mathfrak{S}(N_a)$, it follows from Lemma 1.1.5 and (ii) that $[(PZ)N_a]^o = N_a$; thus by Theorem 1.2.4 $Q(D) = \mathcal{H}$.

From this theorem we get immediately the following result:

<u>Theorem 1.2.7</u>. Let $\mathcal{H}$ be a real Hilbert space and let $\mathfrak{h} = (Z,a)$ be an a.H.n. over $\mathcal{H}$. Assume that

(i*) $D \supset N_a$,

(ii*) There exist constants $c > 0$ and $p > 1$ such that for any $x_1, x_2 \in N_a$

and any $y_1 \in Zx_1$, $y_2 \in Zx_2$ we have

$$\langle y_1-y_2, x_1-x_2\rangle \geq c\|x_1-x_2\|^p, \qquad (1.2.29)$$

(iii*) PZ is maximal monotone on N_a.

Then, for any $e \in \mathcal{H}$, $\mathfrak{h}$ possesses a unique solution corresponding to e.

Proof: If a solution i exists for some $e \in \mathcal{H}$, then due to (1.2.29) and Theorem 1.2.5, i is unique. Thus, we have to show only that $Q(D) = \mathcal{H}$. Referring to Theorem 1.2.6 it will be sufficient to prove that $PZ : N_a \to N_a$ is coercive. To show it, observe that (1.2.29) yields

$$\langle Py_1-Py_2, x_1-x_2\rangle \geq c\|x_1-x_2\|^p$$

for all $x_j \in N_a$, $y_j \in Zx_j$, $j = 1,2$. This is equivalent to

$$\langle z_1-z_2, x_1-x_2\rangle \geq c\|x_1-x_2\|^p \qquad (1.2.30)$$

for all $x_j \in N_a$, $z_j \in PZx_j$, $j = 1,2$.

Now choose $x \in N_a$, $z \in PZx$ and $z_o \in PZ0$; then (1.2.30) implies that

$$\langle z,x\rangle \geq c\|x\|^p + \langle z_o,x\rangle.$$

However, $|\langle z_o,x\rangle| \leq \|z_o\|\cdot\|x\|$, so that $\langle z_o,x\rangle \geq -\|z_o\|\cdot\|x\|$; consequently,

$$\langle z,x\rangle \geq c\|x\|^p - \|z_o\|\cdot\|x\|.$$

Consider the function $\varphi(\xi) = c\xi^p - \|z_o\|\xi$; it is easy to see that $\varphi(\xi) \geq \frac{1}{2}c\xi^p$ for all $\xi \geq c^{-\frac{1}{p-1}}2^{\frac{1}{p-1}}\|z_o\|^{\frac{1}{p-1}} = \xi_o$. Thus, if $\|x\| \geq a \geq \xi_o$, (1.2.31) yields

$$\langle z,x\rangle \geq \tfrac{1}{2}c\|x\|^p \geq \tfrac{1}{2}ca^p;$$

hence, for $a \geq \xi_o$, $q(a) = \inf\{\langle z,x\rangle : z \in PZx,\ x \in N_a,\ \|x\| \geq a\} \geq \frac{1}{2}ca^p$.

Consequently, $\lim_{a\to\infty} a^{-1}q(a) \geq \lim_{a\to\infty} \frac{1}{2}ca^{p-1} = \infty$, i.e., PZ is coercive on N_a.

Remark 4. Inequality (1.2.29) implies immediately that PZ is monotone on N_a; thus, (iii*) requires essentially only the maximality of PZ.

Problems

1.2.1. Let the abstract network $\mathfrak{h} = [Z,N,M]$ over $\mathcal{L}$ be specified by the following data: $\mathcal{L} = R^2$, $N = \{(t,0) : t \in R^1\}$, $M = \{(t,t) : t \in R^1\}$, $D = R^2$, $Z(t,\tau) = (t^3-\tau, -t+\tau^2)$. Show that $Q(D) = R^2$ and that $\mathfrak{h}$ is regular on R^2.

Moreover, find an expression for the admittance A of $\mathfrak{n}$.

1.2.2. Assume that, for some $e \in \mathcal{L}$, $i \in \mathcal{L}$ is a unique solution of an a.n. $\mathfrak{n} = [Z,N,M]$ corresponding to e, (i.e., no other element $i' \in D$ is a solution of $\mathfrak{n}$ corresponding to e); is then the element $v \in Zi$ appearing in the definition of a solution determined uniquely?

1.2.3. Let $\mathfrak{n} = [Z,N,M]$ be an a.n. over $\mathcal{L}$; assume that D is a linear subspace of $\mathcal{L}$ and Z is a linear operator on D. Show that

(i) $Q(D)$ is a linear subspace of $\mathcal{L}$,

(ii) if $\mathfrak{n}$ is regular on D, then the admittance $A : Q(D) \to D$ is a linear operator,

(iii) $\mathfrak{n}$ is regular on $D \Leftrightarrow N \cap R = \{0\}$, where $R = \{\alpha : x \in D, Zx \in M\}$.

1.3 Networks with Z an operator defined on the entire space $\mathcal{H}$; quasi-linear and linear networks

In this section we will assume that Z is an operator defined on the entire space $\mathcal{H}$. We could assume only that $D \supset N_a$, but the assumption $D = \mathcal{H}$ entails no loss of generality in view of our previous results, since all properties of a network depend only on the set $N_a \cap D = N_a$. In other words, if Z is defined only on some $D \supset N_a$, we can define $Zx = 0$ for each $x \notin D$, and the situation remains unchanged.

Theorem 1.3.1. Let $\mathfrak{n} = (Z,a)$ be an a.H.n. over $\mathcal{H}$ with Z being an operator defined on $\mathcal{H}$. Assume that there exist constants $c > 0$ and $\lambda > 0$ such that

(i) $\operatorname{Re}\langle Zx_1-Zx_2,x_1-x_2\rangle \geq c\|x_1-x_2\|^2$ for all $x_1,x_2 \in N_a$,

(ii) $PZ \in \mathrm{Lip}_\lambda[N_a,N_a]$.

Then $Q(\mathcal{H}) = \mathcal{H}$ and $\mathfrak{n}$ is regular on $\mathcal{H}$; moreover, the admittance A of $\mathfrak{n}$ is in $\mathrm{Lip}_\gamma[\mathcal{H},\mathcal{H}]$ with $\gamma = (c\lambda)^{-1}(\lambda+\sqrt{\lambda^2-c^2})$.

Proof: First of all, using the Schwartz inequality and the fact that P is the identity operator on N_a, we conclude easily from (i) and (ii) that $\lambda \geq c$. Next, choose arbitrarily $y \in N_a$ and define the operator $R_y : N_a \to N_a$ by

$$R_y x = r^{-1}(rI-PZ)x + r^{-1}y, \tag{1.3.1}$$

where $r = \lambda^2 c^{-1}$. If $x_1,x_2 \in N_a$, we have by (i) and (ii),

$$\|R_y x_1-R_y x_2\|^2 = r^{-2}\|r(x_1-x_2)-P(Zx_1-Zx_2)\|^2 \tag{1.3.2}$$

$$= r^{-2}\{r^2\|x_1-x_2\|^2 + \|P(Zx_1-Zx_2)\|^2$$
$$- r\langle x_1-x_2, P(Zx_1-Zx_2)\rangle - r\langle P(Zx_1-Zx_2), x_1-x_2\rangle\}$$
$$= r^{-2}\{r^2\|x_1-x_2\|^2 + \|P(Zx_1-Zx_2)\|^2$$
$$- 2r\,\mathrm{Re}\langle P(Zx_1-Zx_2), x_1-x_2\rangle\} \le r^{-2}\{r^2+\lambda^2-2rc\}\|x_1-x_2\|^2$$
$$= u\|x_1-x_2\|^2,$$

where $0 \le u = 1-c^2\lambda^{-2} < 1$. Hence, by (1.3.2), R_y is a contraction mapping on N_a. Consequently, for any $y \in N_a$ there exists a unique $x \in N_a$ such that $x = R_y x$, i.e., $PZx = y$. Thus PZ is 1-to-1 from N_a onto N_a.

Now, since $(PZ)N_a = N_a$, it follows by Theorem 1.2.4 that $Q(\mathcal{H}) = \mathcal{H}$; furthermore, by Theorem 1.2.2, $\mathfrak{h}$ is regular on $\mathcal{H}$.

To conclude the proof, let $y_j \in N_a$, $j = 1,2$ and put $x_j = (PZ)^{-1}y_j$. Then $x_j = R_{y_j}x_j$ and we have by (1.3.1) and (1.3.2),

$$\|x_1-x_2\| = \|R_{y_1}x_1-R_{y_2}x_2\| \le \|R_{y_1}x_1-R_{y_1}x_2\| + \|R_{y_1}x_2-R_{y_2}x_2\|$$
$$\le \sqrt{u}\,\|x_1-x_2\| + r^{-1}\|y_1-y_2\|;$$

hence,

$$\|(PZ)^{-1}y_1 - (PZ)^{-1}y_2\| \le (1-\sqrt{u})^{-1}r^{-1}\|y_1-y_2\|.$$

Thus, $(PZ)^{-1}$ and, consequently, $A = (PZ)^{-1}P$ belongs to $\mathrm{Lip}_\gamma[\mathcal{H},\mathcal{H}]$, where $\gamma = (1-\sqrt{u})^{-1}r^{-1} = (c\lambda)^{-1}(\lambda+\sqrt{\lambda^2-c^2})$.

Remark 1. It is clear that conditions (i), (ii) in Theorem 1.3.1 may be replaced by the following stronger requirements (i)* and (ii)*:

(i)* Inequality (i) holds for all $x_1,x_2 \in \mathcal{H}$,

(ii)* $Z \in \mathrm{Lip}_\lambda[\mathcal{H},\mathcal{H}]$.

The elementary proof of Theorem 1.3.1 relies solely on the contraction mapping principle, i.e., it does not make any use of the powerful Lemma 1.1.5, although our claim follows from theorems in section 1.2. Actually, we can improve our result substantially.

Theorem 1.3.2. Let $\mathfrak{h} = (Z,a)$ be an a.H.n. over a real Hilbert space $\mathcal{H}$, and let Z be an operator defined on $\mathcal{H}$. Assume that

(i)' there exist constants $c > 0$ and $p > 1$ such that

$$\langle Zx_1-Zx_2, x_1-x_2\rangle \ge c\|x_1-x_2\|^p \qquad (1.3.3)$$

for all $x_1, x_2 \in N_a$,

(ii)' PZ is continuous on N_a.

Then $Q(\mathcal{H}) = \mathcal{H}$ and $\mathfrak{h}$ is regular on $\mathcal{H}$; moreover, the admittance A of $\mathfrak{h}$ satisfies the inequality (1.2.8).

Proof: Since $Px = x$ for each $x \in N_a$, inequality (1.3.3) shows that the operator PZ is monotone on N_a. Since PZ is continous on N_a, it follows by Lemma 1.1.6 that $PZ : N_a \to N_a$ is regular on $\mathcal{H}$. The validity of inequality (1.2.8) follows from (c) in Theorem 1.2.5.

Remark 2. As before, conditions (i)', (ii)' can be replaced by the above (i)* and by the requirement that Z is continuous on $\mathcal{H}$. Note that if these conditions are satisfied, we always have regularity and $Q(\mathcal{H}) = \mathcal{H}$, whatever a may be.

Let us now consider the case of a quasi-linear and linear network. To this end we will need the following result:

Lemma 1.3.1. Let $\mathcal{H}_1, \mathcal{H}_2$ be Hilbert spaces and let $A \in [\mathcal{H}_1, \mathcal{H}_2]$. Then we have:

(a) The inverse $A^{-1} : \mathcal{H}_2 \to \mathcal{H}_1$ exists, if and only if

$$\inf_{\substack{\|x\|=1 \\ x\in\mathcal{H}_1}} \sup_{\substack{\|y\|=1 \\ y\in\mathcal{H}_2}} |\langle Ax,y\rangle_2| = c_1 > 0, \tag{1.3.4}$$

and

$$\inf_{\substack{\|y\|=1 \\ y\in\mathcal{H}_2}} \sup_{\substack{\|x\|=1 \\ x\in\mathcal{H}_1}} |\langle Ax,y\rangle_2| = c_2 > 0. \tag{1.3.5}$$

In this case, $A^{-1} \in [\mathcal{H}_2, \mathcal{H}_1]$ and

$$\|A^{-1}\| \le c_1^{-1}. \tag{1.3.6}$$

(b) If $\mathcal{H}_1 = \mathcal{H}_2 = \mathcal{H}$ and there exists $c > 0$ such that

$$|\langle Ax,x\rangle| \ge c\|x\|^2 \tag{1.3.7}$$

for all $x \in \mathcal{H}$, then (1.3.4) and (1.3.5) are satisfied with $c = c_1 = c_2$.

Proof: (a) Assume first that (1.3.4) and (1.3.5) are satisfied. Since

$\sup_{\substack{\|y\|=1 \\ y\in\mathcal{H}_2}} |\langle Ax,y\rangle_2| = \|Ax\|$ for any $x \in \mathcal{H}_1$, (1.3.4) implies that

$$\|Ax\| \geq c_1\|x\| \tag{1.3.8}$$

for every $x \in \mathcal{H}_1$. However, (1.3.8) shows that A is 1-to-1. Next, we are going to show that $A\mathcal{H}_1$ is closed in $\mathcal{H}_2$. Indeed, let $y_n \in A\mathcal{H}_1$, $y_n \to y$; then there exist (uniquely determined) elements $x_n \in \mathcal{H}_1$ such that $y_n = Ax_n$. Since y_n is Cauchy, x_n is also Cauchy according to (1.3.8); thus there exists $x \in \mathcal{H}_1$ such that $x_n \to x$. Consequently, by continuity of A, $Ax_n = y_n \to Ax \in A\mathcal{H}_1$; hence, $y = Ax$ and the closedness of $A\mathcal{H}_1$ is proved.

Now, let us show that $A\mathcal{H}_1 = \mathcal{H}_2$. Indeed, suppose that $A\mathcal{H}_1 \neq \mathcal{H}_2$; then, by a standard theorem [1], there exists $y_o \in \mathcal{H}_2$ with $\|y_o\| = 1$ such that $y_o \perp A\mathcal{H}_1$. Thus, for all $x \in \mathcal{H}_1$ with $\|x\| = 1$, $\langle Ax_1,y_o\rangle_2 = 0$, and consequently, $\sup_{\substack{\|x\|=1 \\ x\in\mathcal{H}_1}} |\langle Ax,y_o\rangle_2| = 0$. Hence, $\inf_{\substack{\|y\|=1 \\ y\in\mathcal{H}_2}} \sup_{\substack{\|x\|=1 \\ x\in\mathcal{H}_1}} |\langle Ax,y\rangle_2| \leq 0$, which contradicts (1.3.5).

Thus, A is 1-to-1 from $\mathcal{H}_1$ onto $\mathcal{H}_2$; the boundedness of A^{-1} and (1.3.6) is a trivial consequence of (1.3.8).

Conversely, assume that A possesses an inverse $A^{-1} : \mathcal{H}_2 \to \mathcal{H}_1$. Then, by the open mapping theorem, $A^{-1} \in [\mathcal{H}_2,\mathcal{H}_1]$. On the other hand, A and A^{-1} possess (uniquely determined) adjoints A^* and $(A^{-1})^*$, respectively, and $A^* \in [\mathcal{H}_2,\mathcal{H}_1]$, $(A^{-1})^* \in [\mathcal{H}_1,\mathcal{H}_2]$ (see [3], p. 131). However, it is easy to see that A^* possesses an inverse and $(A^*)^{-1} = (A^{-1})^*$. Indeed, for any $x \in \mathcal{H}_1$, $y \in \mathcal{H}_2$ we have

$$(*) \qquad \langle (A^{-1})^*x,y\rangle_2 = \langle x,A^{-1}y\rangle_1 .$$

Let $u \in \mathcal{H}_2$ be arbitrary, and put $x = A^*u \in \mathcal{H}_1$; then

$$\langle (A^{-1})^*A^*u,y\rangle_2 = \langle A^*u,A^{-1}y\rangle_1 = \langle u,y\rangle_2,$$

i.e.,

$$\langle [(A^{-1})^*A^*-I]u,y\rangle_2 = 0$$

for any $y \in \mathcal{H}_2$, which implies that $(A^{-1})^*A^* = I$. Hence $(A^{-1})^*$ is a left inverse for A^*.

Analogously, choose $v \in \mathcal{H}_1$ and put $y = Av \in \mathcal{H}_2$ into (*); then

$$\langle (A^{-1})^*x,Av\rangle_2 = \langle A^*(A^{-1})^*x,v\rangle_1 = \langle x,v\rangle_1,$$

which implies that $A^*(A^{-1})^* = I$. Hence, $(A^{-1})^*$ is also a right inverse for A^*. Consequently, $(A^{-1})^* = (A^*)^{-1}$ and our claim is verified.

Next, choose $x \in \mathcal{H}_1$ with $\|x\| = 1$; then

$$\sup_{\substack{\|y\|=1\\ \mathcal{H}_2}} |\langle Ax,y\rangle_2| \geq |\langle Ax, \frac{(A^*)^{-1}x}{\|(A^*)^{-1}x\|}\rangle_2| = \frac{1}{\|(A^*)^{-1}x\|}|\langle x, A^*(A^*)^{-1}x\rangle_1|$$

$$= \|(A^*)^{-1}\|^{-1}\|x\|^{-1} = \|(A^*)^{-1}\|^{-1}.$$

Hence, $\inf\limits_{\substack{\|x\|=1\\ \mathcal{H}_1}} \sup\limits_{\substack{\|y\|=1\\ \mathcal{H}_2}} |\langle Ax,y\rangle_2| \geq \|(A^*)^{-1}\|^{-1}$, i.e., (1.3.4) holds.

Similarly, let $y \in \mathcal{H}_2$ and $\|y\| = 1$; then

$$\sup_{\substack{\|x\|=1\\ \mathcal{H}_1}} |\langle Ax,y\rangle_2| \geq |\langle A \frac{A^{-1}y}{\|A^{-1}y\|}, y\rangle_2| = \frac{1}{\|A^{-1}y\|} \geq \|A^{-1}\|^{-1}.$$

Consequently, $\inf\limits_{\substack{\|y\|=1\\ \mathcal{H}_2}} \sup\limits_{\substack{\|x\|=1\\ \mathcal{H}_1}} |\langle Ax,y\rangle_2| \geq \|A^{-1}\|^{-1}$, so that (1.3.5) holds, too.

(b) Let $x \in \mathcal{H}$ be such that $\|x\| = 1$; then $\sup\limits_{\|y\|=1} |\langle Ax,y\rangle| \geq |\langle Ax,x\rangle| \geq c$, and consequently, $\inf\limits_{\|x\|=1} \sup\limits_{\|y\|=1} |\ldots| \geq c$, i.e., (1.3.4) holds with $c_1 = c$. Similarly, let $y \in \mathcal{H}$ be such that $\|y\| = 1$; then $\sup\limits_{\|x\|=1} |\langle Ax,y\rangle| \geq |\langle Ay,y\rangle| \geq c$. Hence, $\inf\limits_{\|y\|=1} \sup\limits_{\|x\|=1} |\ldots| \geq c$, i.e., (1.3.5) holds, too.

Lemma 1.3.2. If $\mathcal{H}$ is a real Hilbert space and $A \in [\mathcal{H},\mathcal{H}]$, then (1.3.7) holds, if and only if either

$$\langle Ax,x\rangle \geq c\|x\|^2 \text{ for all } x \in \mathcal{H}, \text{ or}$$

$$-\langle Ax,x\rangle \geq c\|x\|^2 \text{ for all } x \in \mathcal{H}.$$

The proof is obvious and is omitted.

Theorem 1.3.3. Let $\mathfrak{n} = (Z,a)$ be an a.H.n. over $\mathcal{H}$ and let Z be an operator defined on $\mathcal{H}$. Assume that $Z = Z_1+Z_2$, where $Z_1 \in [\mathcal{H},\mathcal{H}]$. If

(i) $\inf\limits_{\substack{\|x\|=1\\ x\in N_a}} \sup\limits_{\substack{\|y\|=1\\ y\in N_a}} |\langle Z_1x,y\rangle| = c > 0$, $\inf\limits_{\substack{\|y\|=1\\ y\in N_a}} \sup\limits_{\substack{\|x\|=1\\ x\in N_a}} |\langle Z_1x,y\rangle| > 0$,

(ii) there exists $\lambda > 0$ such that $PZ_2 \in Lip_\lambda\{N_a,N_a\}$,

(iii) $c^{-1}\lambda < 1$,

then $Q(\mathcal{H}) = \mathcal{H}$, $\mathfrak{n}$ is regular on $\mathcal{H}$ and the admittance A of $\mathfrak{n}$ is in $Lip_\delta[\mathcal{H},\mathcal{H}]$, where $\delta = (c-\lambda)^{-1}$.

Proof: First, consider the operator $PZ_1 : N_a \to N_a$; note that N_a, as a closed subspace of $\mathcal{H}$, is a Hilbert space of its own right. If $x,y \in N_a$, we have $\langle PZ_1x,y\rangle = \langle Z_1x,Py\rangle = \langle Z_1x,y\rangle$; hence, conditions (i) show by Lemma 1.3.1 that $(PZ_1)^{-1} : N_a \to N_a$ exists, is bounded and $\|(PZ_1)^{-1}\| \le c^{-1}$.

Next, choose a $y \in N_a$ and define the operator $Q_y : N_a \to N_a$ by

$$Q_yx = -(PZ_1)^{-1}PZ_2x + (PZ_1)^{-1}y. \tag{1.3.9}$$

If $x_1,x_2 \in N_a$, we have by (ii) and (iii)

$$\begin{aligned}\|Q_yx_1-Q_yx_2\| &= \|(PZ_1)^{-1}(PZ_2x_1-PZ_2x_2)\| \qquad (1.3.10)\\ &\le \|(PZ_1)^{-1}\|\cdot\|PZ_2x_1-PZ_2x_2\|\\ &\le u\|x_1-x_2\|\end{aligned}$$

with $u = c^{-1}\lambda < 1$; hence Q_y is a contraction on N_a. Consequently, there exists a unique $x \in N_a$ such that $x = Q_yx$, i.e., in view of (1.3.9), $PZ_1x = -PZ_2x+y$. Thus, $PZx = y$, i.e., PZ is 1-to-1 from N_a onto N_a. This means that $Q(\mathcal{H}) = \mathcal{H}$ by Theorem 1.2.4, and that $\mathfrak{n}$ is regular on $\mathcal{H}$ by Theorem 1.2.2.

The proof of $A \in Lip_\delta[\mathcal{H},\mathcal{H}]$ follows easily from (1.3.10) by using the same argument as in the proof of Theorem 1.3.1.

Obviously, the theorem just proved deals with a "quasi-linear network", i.e., with a network whose operator Z has a small nonlinear part Z_2 (see (ii) and (iii)) compared to the linear part Z_1.

Let us now consider the case of a linear network in more detail.

Theorem 1.3.4. Let $\mathfrak{n} = (Z,a)$ be an a.H.n. over $\mathcal{H}$ and let $PZ \in [N_a,N_a]$, (or $Z \in [\mathcal{H},\mathcal{H}]$).

(a) $Q(\mathcal{H}) = \mathcal{H}$ and $\mathfrak{n}$ is regular on $\mathcal{H}$, if and only if Z satisfies the conditions (i) in Theorem 1.3.3.

(b) If there exists a $c > 0$ such that

$$|\langle Zx,x\rangle| \ge c\|x\|^2 \tag{1.3.11}$$

for all $x \in N_a$, then $Q(\mathcal{H}) = \mathcal{H}$ and $\mathfrak{n}$ is regular on $\mathcal{H}$.
In either case, the admittance A of $\mathfrak{n}$ is in $[\mathcal{H},\mathcal{H}]$ and $\|A\| \leq c^{-1}$.

Proof: (a) Assume that conditions (i) in Theorem 1.3.4 are met. Then, as in the proof of Theorem 1.3.3, the operator PZ is 1-to-1 from N_a onto N_a, and $\|(PZ)^{-1}\| \leq c^{-1}$. Hence, by Theorem 1.2.4, $Q(\mathcal{H}) = \mathcal{H}$, and $\mathfrak{n}$ is regular on $\mathcal{H}$ by Theorem 1.2.2. Also, $\|A\| = \|(PZ)^{-1}P\| \leq \|(PZ)^{-1}\|\cdot\|P\| \leq c^{-1}$.

Conversely, let $Q(\mathcal{H}) = \mathcal{H}$ and $\mathfrak{n}$ be regular on $\mathcal{H}$; then, again by Theorem 1.2.4 and Theorem 1.2.2, $(PZ)N_a = N_a$ and PZ is 1-to-1 on N_a. Hence, by Lemma 1.3.1, PZ satisfies conditions (i) in Theorem 1.3.3, and since $\langle PZx,y\rangle = \langle Zx,y\rangle$ for all $x,y \in N_a$, Z itself satisfies (i).

The proof of (b) is obvious.

Remark 3. If (1.3.11) is satisfied even for all $x \in \mathcal{H}$, then clearly $Q(\mathcal{H}) = \mathcal{H}$ and $\mathfrak{n}$ is regular on $\mathcal{H}$, independently of what the operator a may be.

Problems.

1.3.1 Let $\mathfrak{n} = (Z,a)$ be an a.H.n. over $\mathcal{H}$; assume that

(i) $D = \mathcal{H}$ and $Z \in [\mathcal{H},\mathcal{H}]$,

(ii) $\mathfrak{n}$ is regular on $\mathcal{H}$ and $Q(\mathcal{H}) = \mathcal{H}$.

Then the admittance A of $\mathfrak{n}$ is in $[\mathcal{H},\mathcal{H}]$.

1.3.2 Let $\mathfrak{n} = (Z,a)$ be an a.H.n. over $\mathcal{H}$, and let Z be an operator on $\mathcal{H}$. Assume that

(i) $\langle Zx,x\rangle$ is real and nonnegative for each $x \in \mathcal{H}$,

(ii) $\langle Zx,y\rangle = \langle x,Zy\rangle$ for every $x,y \in \mathcal{H}$.

Then $\mathfrak{n}$ is regular on $\mathcal{H}$, if and only if $N_a \cap N_Z = \{0\}$, where $N_Z = \{x : x \in \mathcal{H}, Zx = 0\}$.

1.3.3 Let $\mathfrak{n} = (Z,a)$ be an a.H.n. over $\mathcal{H}$ and let Z be a linear operator on $\mathcal{H}$. Assume that there exists an integer $n \geq 1$ and $0 < \alpha < 1$ such that $(I+PZ)^n \in Lip_\alpha[N_a,N_a]$, where I denotes the identity operator. Then $\mathfrak{n}$ is regular on $\mathcal{H}$ and $Q(\mathcal{H}) = \mathcal{H}$.
If, in addition, $Z \in [\mathcal{H},\mathcal{H}]$, then the admittance A of $\mathfrak{n}$ is also in $[\mathcal{H},\mathcal{H}]$.

1.3.4 Referring to Theorem 1.3.3 show that the admittance of a quasi-linear a.H.n. can be written as a sum of the admittance of the "linear part" of the network plus a "small" operator, i.e., prove:
Let the assumptions of Theorem 1.3.3 be satisfied, and let A be the admittance of $\mathfrak{n}$; then $A = (PZ_1)^{-1}P + S$, where $S \in Lip_\nu[\mathcal{H},\mathcal{H}]$ with $\nu = \lambda c^{-1}(c-\lambda)^{-1}$.

1.3.5 Show that the solution i of a linear a.H.n. $\mathfrak{n} = (Z,a)$ depends continuously on Z, i.e., prove:

Let $Z_n \in [\mathcal{H},\mathcal{H}]$, $n = 0,1,2,\ldots$ be such that, for any $x \in \mathcal{H}$, $Z_n x \to Z_o x$ in $\mathcal{H}$ as $n \to \infty$. Moreover, assume that there exists a constant $a > 0$ such that, for a fixed $a \in [\mathcal{H},\mathcal{H}']$,

$$\inf_{\substack{\|x\|=1\\ N_a}} \sup_{\substack{\|y\|=1\\ N_a}} |\langle Z_n x,y\rangle| \geq \alpha, \quad \inf_{\substack{\|y\|=1\\ N_a}} \sup_{\substack{\|x\|=1\\ N_a}} |\langle Z_n x,y\rangle| > 0$$

for every $n = 0,1,2,\ldots$. If $e \in \mathcal{H}$ and i_n is the solution of $\mathfrak{h}_n = (Z_n,a)$ corresponding to e for $n = 0,1,2,\ldots$, then $i_n \to i_o$ in $\mathcal{H}$ as $n \to \infty$.

1.4 Equivalent definition of a solution

In this section we will establish an equivalent definition of a solution of an a.H.n. and prove further theorems on regularity. We will need the following propositions:

Lemma 1.4.1. Let L be a linear subspace of a Hilbert space $\mathfrak{H}$; then $L^{\perp\perp} = \bar{L}$.

Lemma 1.4.2. Let $\mathcal{H},\mathcal{H}'$ be Hilbert spaces, let $A \in [\mathcal{H},\mathcal{H}']$ and let $A^* \in [\mathcal{H}',\mathcal{H}]$ be the adjoint of A. Then

$$\{x : x \in \mathcal{H},\ Ax = 0\} = (A^*\mathcal{H}')^{\perp}. \tag{1.4.1}$$

For the proof see Berberian [3], p. 71, 133.

Theorem 1.4.1. Let $\mathfrak{h} = (Z,a)$ be an a.H.n. over $\mathcal{H}$ with $Z : D \to \mathfrak{S}(\mathcal{H})$, $a \in [\mathcal{H},\mathcal{H}']$ and let $a^*\mathcal{H}'$ be closed in $\mathcal{H}$.

(i) If, for some $e \in \mathcal{H}$, $\mathfrak{h}$ possesses a solution $i \in N_a \cap D$, then there exist $u \in Zi$ and $v \in \mathcal{H}'$ such that

$$u + a^*v = e. \tag{1.4.2}$$

(ii) If, for some $e \in \mathcal{H}$, there exist elements $i \in N_a \cap D$, $u \in Zi$ and $v \in \mathcal{H}'$ such that (1.4.2) holds, then i is a solution of $\mathfrak{h}$ corresponding to e.

Proof: By Lemma 1.4.2, $N_a = (a^*\mathcal{H}')^{\perp}$, so that, by our hypothesis and Lemma 1.4.1,

$$N_a^{\perp} = (a^*\mathcal{H}')^{\perp\perp} = \overline{(a^*\mathcal{H}')} = a^*\mathcal{H}'. \tag{1.4.3}$$

(i) Assume that $i \in N_a \cap D$ is a solution of $\mathfrak{h}$ corresponding to $e \in \mathcal{H}$; then, by K_1^*, there exists $u \in Zi$ such that $u - e \in N_a^{\perp}$. Thus, by (1.4.3), there exists element $-v \in \mathcal{H}'$ such that $u-e = a^*(-v)$, i.e. (1.4.2) holds.

(ii) Conversely, if there exist $i \in N_a \cap D$, $u \in Zi$ and $v \in \mathcal{H}'$ such that (1.4.2) holds, we have $u-e = a^*(-v) \in a^*\mathcal{H}' = N_a^{\perp}$; hence, i is a solution of $\mathfrak{h}$ corresponding to $e \in \mathcal{H}$ by K_1^*.

Theorem 1.4.2. Let $\mathfrak{h} = (Z,a)$ be an a.H.n. over $\mathcal{H}$, and let $a^*\mathcal{H}'$ be closed in $\mathcal{H}$. Define the set mapping $K : \Omega = (N_a \cap D) \times N_{a*}^{\perp} \to \mathfrak{S}(\mathcal{H})$, where $N_{a*} = \{y : y \in \mathcal{H}', a^*y = 0\}$, by

$$K(x,y) = Zx + a^*y. \tag{1.4.4}$$

(i) If K is simple on Ω, then $\mathfrak{h}$ is regular on D.

(ii) If Z is an operator and $\mathfrak{h}$ is regular on D, then K is 1-to-1 on Ω.

Proof: (i) Let K be simple on Ω. Suppose that there exists an $e \in Q(D)$ such that $i_1, i_2 \in N_a \cap D$, $i_1 \neq i_2$ are solutions of $\mathfrak{h}$ corresponding to e. Then, by previous theorem, there exist $u_j \in Zi_j$ and $v_j \in \mathcal{H}'$, $j = 1,2$, such that

$$u_1 + a^*v_1 = e, \quad u_2 + a^*v_2 = e. \tag{1.4.5}$$

Also, $u_1 + a^*v_1 \in Zi_1 + a^*v_1 = K(i_1,v_1)$, and similarly $u_2 + a^*v_2 \in K(i_2,v_2)$. By (1.4.5), however, $K(i_1,v_1) \cap K(i_2,v_2) \neq \emptyset$, which implies by simplicity of K that $(i_1,v_1) = (i_2,v_2)$, i.e., $i_1 = i_2$ - contradiction. Thus, K simple on $\Omega \Rightarrow \mathfrak{h}$ regular on D.

(ii) Let Z be an operator and assume that $\mathfrak{h}$ is regular on D. In this case, K is an operator. Suppose that K is not 1-to-1, i.e., there exists $(i_j,v_j) \in \Omega$, $j = 1,2$, $(i_1,v_1) \neq (i_2,v_2)$ such that $e = K(i_1,v_1) = K(i_2,v_2)$. Then, by (1.4.4),

$$Zi_1 + a^*v_1 = e, \quad Zi_2 + a^*v_2 = e. \tag{1.4.6}$$

Since $i_1, i_2 \in N_a \cap D$, it follows from (1.4.6) by previous theorem that both i_1 and i_2 are solutions of $\mathfrak{h}$ corresponding to e, and consequently, by regularity of $\mathfrak{h}$, $i_1 = i_2$. Hence, by (1.4.6), $a^*(v_1-v_2) = 0$, i.e., $v_1-v_2 \in N_{a*}$. However, since $v_1, v_2 \in N_{a*}^{\perp}$, it follows that $v_1-v_2 = 0$. Thus, $(i_1,v_1) = (i_2,v_2)$ - contradiction. Hence, K is 1-to-1, what we wanted to prove.

Remark 1. The assertion (ii) need not hold, if the assumption that Z be an operator is dropped. Indeed, let $\mathfrak{h}$ be regular on D, and suppose that K is not simple, i.e., there exists $(i_j,v_j) \in \Omega$, $j = 1,2$, $(i_1,v_1) \neq (i_2,v_2)$ such that $M = K(i_1,v_1) \cap K(i_2,v_2) \neq \emptyset$. If $e \in M$, then by (1.4.4) there exists $u_j \in Zi_j$, $j = 1,2$, such that $u_1 + a^*v_1 = e$ and $u_2 + a^*v_2 = e$. By Theorem 1.4.1 it follows that i_1 and i_2 are solutions of h corresponding to e, so that by regularity, $i_1 = i_2$. However, now we get only $u_1 - u_2 = a^*(v_2 - v_1)$, from which we cannot conclude that $v_1 = v_2$ and thus get a contradiction.

The situation simplifies, if Z is an operator defined on the entire space $\mathcal{H}$. Indeed, we have the following result:

Theorem 1.4.3. Let $h = (Z,a)$ be an a.H.n. over $\mathcal{H}$ with Z being an operator defined on $\mathcal{H}$, and let $a^*\mathcal{H}'$ be closed in $\mathcal{H}$. Define the operator $K' : N_a \times N_{a^*}^{\perp} \to \mathcal{H}$ by

$$K'(x,y) = Zx + a^*y. \tag{1.4.7}$$

Then

(i) $Q(\mathcal{H}) = \mathcal{H} \Leftrightarrow K'$ is onto $\mathcal{H}$,

(ii) h is regular on $\mathcal{H} \Leftrightarrow K'$ is 1-to-1 on $N_a \times N_{a^*}^{\perp}$.

Proof: The proposition (ii) follows immediately from Theorem 1.4.2. To prove (i), assume first that K' is onto $\mathcal{H}$. This means that, for any $e \in \mathcal{H}$, there exists $(i,v) \in N_a \times N_{a^*}^{\perp}$ such that $K'(i,v) = e$, i.e., $Zi + a^*y = e$. Since $i \in N_a$ and $y \in N_{a^*}^{\perp} \subset \mathcal{H}'$, Theorem 1.4.1 (ii) implies that i is a solution of h corresponding to e. Hence, $Q(\mathcal{H}) = \mathcal{H}$.

Conversely, assume that $Q(\mathcal{H}) = \mathcal{H}$. Choose $e \in \mathcal{H}$; then there exists $i \in N_a$ which is a solution of h corresponding to e. By Theorem 1.4.1 (i) however, there exist $v \in \mathcal{H}'$ such that $Zi + a^*v = e$. Since $\mathcal{H}' = N_{a^*} \oplus N_{a^*}^{\perp}$, there exist (even unique) elements $v' \in N_{a^*}$ and $v'' \in N_{a^*}^{\perp}$ such that $v = v' + v''$. However, $a^*v = a^*v' + a^*v'' = a^*v''$, so that the above equation yields $Zi + a^*v'' = e$, i.e., by (1.4.7), $K'(i,v'') = e$ and $(i,v'') \in N_a \times N_{a^*}^{\perp}$. Hence, K' is onto $\mathcal{H}$ which concludes the proof.

Another modification of the above results is obtained, if Z is a linear operator defined on $\mathcal{H}$. Indeed, we have the result:

Theorem 1.4.4. Let $h = (Z,a)$ be an a.H.n. over $\mathcal{H}$ with Z being a linear operator defined on $\mathcal{H}$, and let $a^*\mathcal{H}'$ be closed in $\mathcal{H}$. Define the operator $M : \mathcal{H} \times N_{a^*}^{\perp} \to \mathcal{H} \times a\mathcal{H}$ by

$$M(x,y) = (Zx + a^*y, ax). \tag{1.4.8}$$

Then $Q(\mathcal{H}) = \mathcal{H}$ and h is regular on $\mathcal{H}$ if and only if M possesses an inverse M^{-1}.

Proof: Assume first that M possesses an inverse. Choose $e \in \mathcal{H}$; then there exists a unique pair $(i,v) \in \mathcal{H} \times N_{a*}^{\perp}$ such that $M(i,v) = (e,0)$, i.e., $Zi + a^*v = e$ and $ai = 0$. This means by Theorem 1.4.1 (ii) (note that $N_{a*}^{\perp} \subset \mathcal{H}'$) that i is a solution of $\mathfrak{n}$ corresponding to e. Hence, $Q(\mathcal{H}) = \mathcal{H}$.

Furthermore, suppose that the above element $e \in \mathcal{H}$ is such that $\mathfrak{n}$ possesses solutions i_1, i_2 corresponding to e. Then, by Theorem 1.4.1, there exist $v_1, v_2 \in \mathcal{H}'$ such that $Zi_j + a^*v_j = e$, $j = 1,2$. As we have seen above, the v_j's can be selected so that $v_j \in N_{a*}^{\perp}$. Since $i_j \in N_a$, we have $(i_j, v_j) \in \mathcal{H} \times N_{a*}^{\perp}$ and $M(i_j, v_j) = (e,0)$, $j = 1,2$; thus, by the existence of M^{-1}, $(i_1,v_1) = (i_2,v_2) \Rightarrow i_1 = i_2$. Hence, $\mathfrak{n}$ is regular on $\mathcal{H}$.

Conversely, assume that $Q(\mathcal{H}) = \mathcal{H}$ and $\mathfrak{n}$ is regular on $\mathcal{H}$. Choose arbitrarily $(\xi,\eta) \in \mathcal{H} \times a\mathcal{H}$. Then there exists at least one $x_o \in \mathcal{H}$ such that $ax_o = \eta$. Putting $e = \xi - Zx_o \in \mathcal{H}$ it follows from our hypothesis $Q(\mathcal{H}) = \mathcal{H}$ that there exists an $x \in \mathcal{H}$ which is a solution of $\mathfrak{n}$ corresponding to e. Thus, by Theorem 1.4.1 (i), there exists $y \in \mathcal{H}'$ such that $Zx + a^*y = \xi - Zx_o = e$. Selecting y so that $y \in N_{a*}^{\perp}$, we have $Z(z+x_o) + a^*y = \xi$ and $a(x+x_o) = ax_o = \eta$, i.e., $(x+x_o, y) \in \mathcal{H} \times N_{a*}^{\perp}$ and $M(x+x_o, y) = (\xi,\eta)$. Hence, the operator M is onto $\mathcal{H} \times a\mathcal{H}$.

Next, noting that M is linear, assume that, for some pair $(x,y) \in \mathcal{H} \times N_{a*}^{\perp}$, $M(x,y) = 0$. This means that $Zx + a^*y = 0$ and $ax = 0$, i.e., by Theorem 1.4.1 (ii), x is a solution of $\mathfrak{n}$ corresponding to $e = 0$. By regularity of $\mathfrak{n}$, $x = 0$, and consequently, by the above equation, $a^*y = 0$. Since $y \in N_{a*}^{\perp}$, it follows that $y = 0$. Hence, $M(x,y) = 0 \Rightarrow (x,y) = 0$, i.e., M is 1-to-1.

Problems

1.4.1 Referring to Theorem 1.4.4, prove the proposition: the operator M defined by (1.4.8) is bounded, if and only if Z is bounded. (We understand that $\mathcal{H} \times \mathcal{H}'$ is equipped with the inner product $\langle (x_1,x_1'),(x_2,x_2')\rangle = \langle x_1,x_2\rangle + \langle x_1',x_2'\rangle'$ for all $(x_j,x_j') \in \mathcal{H} \times \mathcal{H}'$, $j = 1,2$).

1.5 Qualitative properties of abstract networks

In this section we introduce a generalized concept of monotonicity (positivity) and show that Z and the admittance A of a network are monotone (positive) simultaneously. Also, we will establish a certain property of the "power" associated with the network, and prove two theorems, which will enable us to show that a solution of an infinite Hilbert network is a limit of solutions of finite subnetworks.

Let $\varphi : C^1 \to R^1$ be a fixed function with the property $\varphi(\bar{\xi}) = \varphi(\xi)$ for all $\xi \in C^1$. (Here, C^1 signifies the complex plane.) Let $\mathfrak{H}$ be a Hilbert space, let $\tilde{\mathfrak{D}} \subset \mathfrak{H}$, $\tilde{\mathfrak{D}} \neq \emptyset$ and let $M : \tilde{\mathfrak{D}} \to \mathfrak{S}(\mathfrak{H})$. We will write

(a) $M \in \mathfrak{P}_\varphi(\tilde{\mathfrak{D}})$, if $\varphi(\langle y,x\rangle) \geq 0$ for all $x \in \tilde{\mathfrak{D}}$ and all $y \in Mx$,

(b) $M \in \mathfrak{M}_\varphi(\tilde{\mathfrak{D}})$, if $\varphi(\langle y_1-y_2, x_1-x_2\rangle) \geq 0$ for all $x_j \in \tilde{\mathfrak{D}}$, $y_j \in Mx_j$.

For example, we can put $\varphi(\xi) = \mathrm{Re}\ \xi$, or $\varphi(\xi) = \alpha\ \mathrm{Re}\ \xi + \beta|\mathrm{Im}\ \xi|$, α,β real etc.

Theorem 1.5.1. Let $\mathfrak{h} = (Z,a)$ be an a.H.n. over $\mathfrak{H}$, let $N_a \cap D \neq \emptyset$, and let $A : Q(D) \to \mathfrak{S}(N_a \cap D)$ be the admittance of $\mathfrak{h}$; then

(i) $A \in \mathfrak{P}_\varphi(Q(D)) \Leftrightarrow Z \in \mathfrak{P}_\varphi(N_a \cap D)$,

(ii) $A \in \mathfrak{M}_\varphi(Q(D)) \Leftrightarrow Z \in \mathfrak{M}_\varphi(N_a \cap D)$.

Proof: (i) Choose $e \in Q(D)$ and let $i \in Ae$. By K_1^* and K_2^*, $i \in N_a \cap D$ and there exists $v \in Zi$ such that $v-e \in N_a^\perp$; hence, $\langle i,v-e\rangle = 0$, i.e.,

$$\langle i,e\rangle = \overline{\langle v,i\rangle}. \tag{1.5.1}$$

Now, if $Z \in \mathfrak{P}_\varphi(N_a \cap D)$, (1.5.1) shows that $A \in \mathfrak{P}_\varphi(Q(D))$.

Conversely, assume that $A \in \mathfrak{P}_\varphi(Q(D))$; choose arbitrarily $i \in N_a \cap D$ and some $v \in Zi$. Since $v \in [Z(N_a \cap D)]^o$, then $Pv \in [(PZ)(N_a \cap D)]^o \subset Q(D)$; thus, we can set $e = Pv$. Then $P(v-e) = Pv-Pe = Pv - P^2v = Pv - Pv = 0$, i.e., $v-e \in N_a^\perp$. Hence, i is a solution of $\mathfrak{h}$ corresponding to e, and consequently, (1.5.1) holds. Thus, $\varphi(\langle v,i\rangle) \geq 0$, i.e., $Z \in \mathfrak{P}_\varphi(N_a \cap D)$ and (i) is proven.

(ii) Let $e_j \in Q(D)$ and let $i_j \in Ae_j$, $j = 1,2$; then

$$\langle i_1-i_2, e_1-e_2\rangle = \langle i_1,e_1\rangle - \langle i_2,e_1\rangle - \langle i_1,e_2\rangle + \langle i_2,e_2\rangle. \tag{1.5.2}$$

On the other hand, there exist $v_j \in Zi_j$, $j = 1,2$ such that $v_j-e_j \in N_a^\perp$, i.e., $\langle c,v_j\rangle = \langle c,e_j\rangle$ for each $c \in N_a$. However, since $i_j \in N_a$, we get relations

$$\langle i_1,v_1\rangle = \langle i_1,e_1\rangle,\ \langle i_2,v_1\rangle = \langle i_2,e_1\rangle,$$
$$\langle i_1,v_2\rangle = \langle i_1,e_2\rangle,\ \langle i_2,v_2\rangle = \langle i_2,e_2\rangle.$$

Introducing this into (1.5.2), we get readily

$$\langle i_1-i_2, e_1-e_2\rangle = \overline{\langle v_1-v_2, i_1-i_2\rangle}. \tag{1.5.3}$$

The rest of the proof follows the same pattern as in (i).

Corollary: If $A : Q(D) \to \mathfrak{S}(N_a \cap D)$ is the admittance of an a.H.n., then (i)* $Z \in \mathfrak{P}_{\varphi}(D) \Rightarrow A \in \mathfrak{P}_{\varphi}(Q(D))$,

(ii)* $Z \in \mathfrak{M}_{\varphi}(D) \Rightarrow A \in \mathfrak{M}_{\varphi}(Q(D))$.

The proof is obvious.

In Theorem 1.2.4 (c) we have shown that, under the condition (1.2.10), the quantity $\mathrm{Re}\langle Ae_1-Ae_2, e_1-e_2\rangle$ is nonnegative. This estimate can be improved, if we assume that PZ is an operator satisfying Hölder's condition. Indeed, we have the result:

<u>Theorem 1.5.2</u>. Let $\mathfrak{n} = (Z,a)$ be an a.H.n. over $\mathfrak{H}$, and let Z be an operator on $D \subset \mathfrak{H}$. Assume that

(i) there exist $c > 0$ and $p > 1$ such that

$$\mathrm{Re}\langle Zx_1-Zx_2, x_1-x_2\rangle \geq c\|x_1-x_2\|^p \tag{1.5.4}$$

for all $x_1, x_2 \in N_a \cap D$,

(ii) there exist $d > 0$ and $q > 0$ such that

$$\|PZx_1-PZx_2\| \leq d\|x_1-x_2\|^q$$

for all $x_1, x_2 \in N_a \cap D$.

Then $\mathfrak{n}$ is regular on D and the admittance A of $\mathfrak{n}$ satisfies the inequality

$$\mathrm{Re}\langle Ae_1-Ae_2, e_1-e_2\rangle \geq cd^{-\frac{p}{q}}\|Pe_1-Pe_2\|^{\frac{p}{q}} \tag{1.5.6}$$

for all $e_1, e_2 \in Q(D)$.

<u>Proof</u>: From (1.5.5) we have $\|x_1-x_2\| \geq d^{-\frac{1}{q}}\|PZx_1-PZx_2\|^{\frac{1}{q}}$; since $Px = x$ for any $x \in N_a$, (1.5.4) yields for $x_1, x_2 \in N_a \cap D$,

$$\begin{aligned}\mathrm{Re}\langle PZx_1-PZx_2, x_1-x_2\rangle &= \mathrm{Re}\langle Zx_1-Zx_2, P(x_1-x_2)\rangle \\ &\geq c\|x_1-x_2\|^p \geq cd^{-\frac{p}{q}}\|PZx_1-PZx_2\|^{\frac{p}{q}}.\end{aligned} \tag{1.5.7}$$

Now, choose $e_j \in Q(D)$, $j = 1,2$ and put $x_j = Ae_j = (PZ)^{-1}Pe_j \in N_a \cap D$ into (1.5.7); we get

$$\mathrm{Re}\langle Pe_1-Pe_2, Ae_1-Ae_2\rangle \geq cd^{-\frac{p}{q}}\|Pe_1-Pe_2\|^{\frac{p}{q}}.$$

Since $PAe_j = Ae_j$, (1.5.6) follows immediately.

Let us now prove a result which is a basis for proving the "least power theorem".

If $e \in \mathcal{H}$, let

$$F_e = \{x : x \in \mathcal{H}, \operatorname{Re}\langle e,x\rangle = 0\}. \qquad (1.5.9)$$

It is clear that F_e is a real linear subspace of $\mathcal{H}$, i.e., $\alpha x + \beta y \in F_e$ whenever $x,y \in F_e$ and α, β are real; also note that $\{e\}^{\perp} \subset F_e$.

Theorem 1.5.3. Let $\mathfrak{h} = (Z,a)$ be an a.H.n. over $\mathcal{H}$ with Z being a linear operator defined on $\mathcal{H}$. Assume that

(i) $\operatorname{Re}\langle Zx,y\rangle = \operatorname{Re}\langle x,Zy\rangle$ for all $x,y \in N_a$,

(ii) $\operatorname{Re}\langle Zx,x\rangle > 0$ for all $x \in N_a$, $x \neq 0$.

Let $e \in \mathcal{H}$ be such that $F_e \cap N_a \neq \{0\}$, and assume that $i \in N_a$ is a solution of $\mathfrak{h}$ corresponding to e. Let $\rho : i + F_e \cap N_a \to R^1$ be defined by

$$\rho(x) = \operatorname{Re}(\langle Zx,x\rangle - \langle Zi,i\rangle). \qquad (1.5.10)$$

Then

(a) i is determined uniquely,

(b) $\rho(x) > 0$ for every $x \in i + F_e \cap N_a$, $x \neq i$.

Proof: Since Z is linear, (a) follows immediately from Theorem 1.2.5 (a). To prove (b) note that, by K_1^*, K_2^*, $i \in N_a$ and $\langle c,e\rangle = \langle c,Zi\rangle$ for all $c \in N_a$; thus, for any $i' \in F_e \cap N_a$, $\operatorname{Re}\langle i',e\rangle = 0 = \operatorname{Re}\langle i',Zi\rangle$. Hence, by (i),

$$\operatorname{Re}\langle Zi,i'\rangle = 0, \ \operatorname{Re}\langle Zi',i\rangle = 0. \qquad (1.5.11)$$

Now, let $x \in i + F_e \cap N_a$, $x \neq i$; then $x = i+i'$ with $i' \in F_e \cap N_a$, $i' \neq 0$, and we have by (1.5.11) and (ii), $\rho(x) = \operatorname{Re}(\langle Z(i+i'),i+i'\rangle - \langle Zi,i\rangle) = \operatorname{Re}\langle Zi,i\rangle + \operatorname{Re}\langle Zi',i\rangle + \operatorname{Re}\langle Zi,i'\rangle + \operatorname{Re}\langle Zi',i'\rangle - \operatorname{Re}\langle Zi,i\rangle = \operatorname{Re}\langle Zi',i'\rangle > 0$. Hence, the proof.

Remark 1. (a) The assumptions (i), (ii) in Theorem 1.5.3 are clearly satisfied, if Z is a selfadjoint strictly positive operator, i.e., $\langle Zx,x\rangle$ is real and positive for all $x \in \mathcal{H}$, $x \neq 0$.

(b) If $\mathcal{H}$ is a real Hilbert space, then condition (i) automatically implies that $PZ : N_a \to N_a$ is a linear bounded operator. To see this, it suffices to realize that (i) then can be written as $\langle PZx,y\rangle = \langle x,PZy\rangle$ for all $x,y \in N_a$, and make use of the following proposition:

Hellinger's Theorem [4]: Let $\mathfrak{H}$ be a (not necessarily real) Hilbert space, and let $M : \mathfrak{H} \to \mathfrak{H}$ be such that, for all $x,y \in \mathfrak{H}$,

$$(*) \qquad \langle Mx,y\rangle = \langle x,My\rangle;$$

then $M \in [\mathfrak{H},\mathfrak{H}]$.

Indeed, if $x_1,x_2 \in \mathfrak{H}$ and α_1,α_2 are scalars, then (*) yields $\langle M(\alpha_1x_1+\alpha_2x_2),y\rangle = \langle \alpha_1x_1+\alpha_2x_2,My\rangle = \alpha_1\langle x_1,My\rangle + \alpha_2\langle x_2,My\rangle = \langle \alpha_1Mx_1,y\rangle + \langle \alpha_2Mx_2,y\rangle$, i.e., $\langle M(\alpha_1x_1+\alpha_2x_2) - \alpha_1Mx_1-\alpha_2Mx_2,y\rangle = 0$ for any $y \in \mathfrak{H}$. Hence $M(\alpha_1x_1+\alpha_2x_2) - \alpha_1Mx_1 - \alpha_2Mx_2 = 0$, i.e., M is linear.

Next, assume that $x_n \to x$ and $Mx_n \to z$; then, by continuity of the inner product, for any $y \in \mathcal{H}$, $\langle Mx_n,y\rangle \to \langle z,y\rangle$ and $\langle x_n,My\rangle \to \langle x,My\rangle$. However, by (*), $\langle Mx_n,y\rangle = \langle x_n,My\rangle$, and consequently, $\langle z,y\rangle = \langle x,My\rangle = \langle Mx,y\rangle$; thus, since $y \in \mathcal{H}$ is arbitrary, $z = Mx$, i.e., the graph of M is closed. Hence, by the closed graph theorem [1], M is bounded.

Finally, let us prove two theorems that will be used in showing that any solution of an infinite Hilbert network is a limit of solutions of finite subnetworks.

Theorem 1.5.4. Let $\mathfrak{n} = (Z,a)$ be an a.H.n. over $\mathcal{H}$ with Z being an operator defined on $\mathcal{H}$. Let constants $c,\lambda > 0$ exist such that

$$\mathrm{Re}\langle Zx_1-Zx_2,x_1-x_2\rangle \geq c\|x_1-x_2\|^2 \tag{1.5.12}$$

for all $x_1,x_2 \in N_a$, and

$$PZ \in \mathrm{Lip}_\lambda[N_a,N_a]. \tag{1.5.13}$$

Furthermore, assume that the following conditions are satisfied:

(i) $\mathcal{H}_n$, $n = 1,2,3,\ldots$ is a sequence of closed linear subspaces of $\mathcal{H}$ such that $\mathcal{H}_1 \subset \mathcal{H}_2 \subset \mathcal{H}_3 \subset\ldots\subset \mathcal{H}$, and P_n is the orthogonal projection from $\mathcal{H}$ onto $\mathcal{H}_n$.

(ii) For every n, $a_n \in [\mathcal{H},\mathcal{H}']$ is such that

$$N_{a_n} = \{x : x \in \mathcal{H},\ a_nx = 0\} \subset \mathcal{H}_n,$$

and $N_{a_n} \subset N_{a_{n+1}}$.

(iii) $N = \bigcup_{n=1}^{\infty} N_{a_n}$ is dense in N_a, i.e., $\bar{N} = N_a$.

Let $e \in \mathcal{H}$, let $i \in \mathcal{H}$ be the (unique) solution of $\mathfrak{n}$ corresponding to e, and let $i_n \in \mathcal{H}$ be the (unique) solution of $\mathfrak{n}_n = (Z, a_n)$ corresponding to $P_n e$. Then $i_n \to i$ in $\mathcal{H}$.

If, in addition, $\mathfrak{n}$ is linear, then conditions (1.5.12), (1.5.13) may be replaced by requirements

$$|\langle Zx, x\rangle| \geq c\|x\|^2, \; c > 0, \tag{1.5.14}$$

for all $x \in N_a$, and

$$Z \in [\mathcal{H}, \mathcal{H}]. \tag{1.5.15}$$

<u>Proof</u>: First of all, from Theorem 1.3.1 it follows that $\mathfrak{n}$ is regular on $\mathcal{H}$ and $Q(\mathcal{H}) = \mathcal{H}$. Moreover, by (iii), $N_{a_n} \subset N_a$ for every n, and consequently, (1.5.12) holds for all $x_1, x_2 \in N_{a_n}$. Let Q_n be the orthogonal projection from $\mathcal{H}$ onto N_{a_n}; due to inclusion $N_{a_n} \subset N_a$ we have $Q_n = Q_n P$. Consequently, (1.5.13) yields for all $x_1, x_2 \in N_{a_n}$, $\|Q_n(Zx_1 - Zx_2)\| = \|Q_n P(Zx_1 - Zx_2)\| \leq \|P(Zx_1 - Zx_2)\| \leq \lambda\|x_1 - x_2\|$, i.e., $Q_n Z \in \mathrm{Lip}_\lambda[N_{a_n}, N_{a_n}]$; hence each network $\mathfrak{n}_n$ is regular on $\mathcal{H}$ and $Q(\mathcal{H}) = \mathcal{H}$ by Theorem 1.3.1.

Next, choose $e \in \mathcal{H}$ and construct the solution i of $\mathfrak{n}$ corresponding to e. If $\varepsilon > 0$, then by (iii) there exists $j \in N$ such that $\|j - i\| < \frac{\varepsilon}{2}\min[1, c\lambda^{-1}]$. Thus, due to (ii), there exists an integer $M > 0$ so that $j \in N_{a_n}$ for all $n \geq M$. Choose some $n \geq M$, and let i_n be the solution of $\mathfrak{n}_n$ corresponding to $P_n e$; thus, we have $i_n \in N_{a_n}$ and

$$\langle c, Zi_n\rangle = \langle c, P_n e\rangle \tag{1.5.16}$$

for every $c \in N_{a_n}$.

On the other hand, we have $i \in N_a$ and

$$\langle c', Zi\rangle = \langle c', e\rangle \tag{1.5.17}$$

for all $c' \in N_a$, and consequently, also for all $c \in N_{a_n}$. However, if $c \in N_{a_n} \subset \mathcal{H}_n$, we have

$$\langle c, P_n e\rangle = \langle P_n c, e\rangle = \langle c, e\rangle.$$

Hence, (1.5.16), (1.5.17) yield $\langle c, Zi_n - Zi\rangle = 0$, and consequently,

$$\langle c, Zi_n - Zj\rangle = \langle c, Zi - Zj\rangle \tag{1.5.18}$$

for any $c \in N_{a_n}$. Since $i_n, j \in N_{a_n}$, we can put $c = i_n - j$ and get from (1.5.18),

$$\langle i_n - j, Zi_n - Zj\rangle = \langle i_n - j, Zi - Zj\rangle. \tag{1.5.19}$$

Using (1.5.12), (1.5.13) and the fact that P is an identity on N_{a_n}, we obtain from (1.5.19),

$$\begin{aligned} c\|i_n - j\|^2 &\le \mathrm{Re}\langle i_n - j, Zi_n - Zj\rangle \qquad (1.5.20)\\ &= \mathrm{Re}\langle i_n - j, Zi - Zj\rangle \le |\langle P(i_n - j), Zi - Zj\rangle| \\ &= |\langle i_n - j, PZi - PZj\rangle| \le \|i_n - j\|\cdot\|PZi - PZj\| \le \|i_n - j\|\cdot\lambda\|i - j\|. \end{aligned}$$

Consequently, $\|i_n - j\| \le c^{-1}\lambda\|i - j\| < \frac{\varepsilon}{2}$; thus we have finally, $\|i_n - i\| \le \|i_n - j\| + \|j - i\| < \varepsilon$ what we wanted to show.

As for the last proposition concerning a linear network, observe first that $\mathfrak{h}$ and each $\mathfrak{h}_n$ is regular on $\mathcal{H}$ and $Q(\mathcal{H}) = \mathcal{H}$ by Theorem 1.3.3 (b). On the other hand, except for the proof of regularity of $\mathfrak{h}$ and $\mathfrak{h}_n$, (1.5.12) was used only for establishing the inequality (1.5.20). However, it is easy to see that (1.5.20) follows from (1.5.14) and (1.5.15), too; hence the proof.

Theorem 1.5.5. The Theorem 1.5.4 remains true, if $\mathcal{H}$ is a real Hilbert space and the assumptions (1.5.12), (1.5.13) are replaced by the following conditions (i)* and (ii)*:

(i)* there exist constants $c > 0$ and $p \ge 2$ such that

$$\langle Zx_1 - Zx_2, x_1 - x_2\rangle \ge c\|x_1 - x_2\|^p \tag{1.5.21}$$

for all $x_1, x_2 \in N_a$,

(ii)* PZ is continuous on N_a.

Proof: First, referring to Theorem 1.3.2, it follows that $\mathfrak{h}$ is regular on $\mathcal{H}$ and $Q(\mathcal{H}) = \mathcal{H}$. Since $N_{a_n} \subset N_a$ for each n, (1.5.21) holds for all $x_1, x_2 \in N_{a_n}$. Also, if Q_n is the orthogonal projection from $\mathcal{H}$ onto N_{a_n}, we have $Q_n = Q_n P$; thus by (ii)*, the operator $Q_n Z = Q_n(PZ)$ is continuous on N_{a_n}. Consequently, by Theorem 1.3.2, each network $\mathfrak{h}_n$ is regular on $\mathcal{H}$ and $Q(\mathcal{H}) = \mathcal{H}$.

Next, choose $e \in \mathcal{H}$ and construct the solution i of $\mathfrak{h}$ corresponding to e. Let $\varepsilon > 0$. Since PZ is continuous on N_a and $i \in N_a$, there exists $\delta > 0$ such that

$$\|PZi-PZx\| < c(\frac{\varepsilon}{2})^{p-1} \tag{1.5.22}$$

whenever $x \in N_a$ and $\|i-x\| < \delta$.

As in the proof of Theorem 1.5.4, there exists $j \in N$ such that $\|j-i\| < \min[\frac{\varepsilon}{2},\delta]$, and $j \in N_{a_n}$ for all $n \geq M$. Choose some $n \geq M$, and denote i_n the solution of $\mathfrak{h}_n$ corresponding to $P_n e$. Then we conclude in the same manner as before that (1.5.19) holds. However, since $\|j-i\| < \delta$, (1.5.22) is satisfied for $x = j$, and we get from (1.5.21) and (1.5.19),

$$c\|i_n-j\|^p \leq \|i_n-j\|\cdot\|PZi-PZj\| \leq \|i_n-j\|\cdot c(\frac{\varepsilon}{2})^{p-1}. \tag{1.5.23}$$

Hence, by (1.5.23), $\|i_n-j\| \leq \frac{\varepsilon}{2}$. Since $\|j-i\| < \frac{\varepsilon}{2}$, it follows that $\|i_n-i\| < \varepsilon$ which concludes the proof.

1.6 Causality

In this section, we will establish conditions under which the admittance of an a.H.n. is causal. We will use the causality concept introduced by Saeks [5].

For every $T \in R^1$, let S_T be an orthogonal projection of $\mathcal{H}$ onto itself, and let the collection $\{S_T : T \in R^1\}$ be a resolution of identity on $\mathcal{H}$, i.e.,

(i) $S_{T_1} \leq S_{T_2}$ for $T_1 \leq T_2$ (i.e., $\langle (S_{T_2}-S_{T_1})x,x\rangle$ is real and nonnegative for each $x \in \mathcal{H}$),

(ii) for every $T_o \in R^1$ and $x \in \mathcal{H}$, $S_T x \to S_{T_o} x$ as $T \to T_o$, $T > T_o$ (i.e., $S_T x$ is right continuous for each $x \in \mathcal{H}$),

(iii) for every $x \in \mathcal{H}$, $S_T x \to 0$ as $T \to -\infty$, and $S_T x \to x$ as $T \to \infty$.

Definition: Let $D \subset \mathcal{H}$, $D \neq \emptyset$, let $A : D \to \mathfrak{S}(\mathcal{H})$ be a set mapping, and let $\mathfrak{M} \subset D$, $\mathfrak{M} \neq \emptyset$; A will be called causal on $\mathfrak{M}$, if

$$x_1,x_2 \in \mathfrak{M},\ S_T x_1 = S_T x_2 \Rightarrow S_T A x_1 = S_T A x_2. \tag{1.6.1}$$

Lemma 1.6.1. Let $\mathfrak{M} \subset D$ be a nonempty set such that $S_T\mathfrak{M} \subset \mathfrak{M}$ for every $T \in R^1$, and let $A : D \to \mathfrak{S}(\mathcal{H})$; then A is causal on $\mathfrak{M}$, if and only if $S_T A = S_T A S_T$ on $\mathfrak{M}$ for every $T \in R^1$.

Proof: Let A be causal on $\mathfrak{M}$. Choose $x \in \mathfrak{M}$; since $S_T = S_T^2$ and $S_T x \in \mathfrak{M}$, we have $S_T x = S_T(S_T x)$, and consequently, by causality, $S_T A x = S_T A(S_T x)$. Since x is arbitrary (and T, too) it follows that $S_T A = S_T A S_T$ on $\mathfrak{M}$ for every $T \in R^1$.

Conversely, let $S_T A = S_T A S_T$ on $\mathfrak{M}$ for each $T \in R^1$. Assume that

$x_1, x_2 \in \mathfrak{M}$ and $S_T x_1 = S_T x_2$ for some $T \in R^1$. Since $S_T x_1, S_T x_2 \in \mathfrak{M}$, we have $S_T A x_1 = S_T A(S_T x_1)$, and $S_T A x_2 = S_T A(S_T x_2)$; hence, $S_T A x_1 = S_T A x_2$, what we wanted to show.

Theorem 1.6.1. Let $\mathfrak{h} = (Z,a)$ be an a.H.n. over $\mathfrak{H}$. Assume that

$$\langle z_1 - z_2, x_1 - x_2 \rangle \neq 0 \qquad (1.6.2)$$

for all $x_1, x_2 \in N_a \cap D$, $x_1 \neq x_2$ and all $z_1 \in Zx_1$, $z_2 \in Zx_2$. Moreover, assume that

(i) for each $T \in R^1$, the projection S_T commutes with P,

(ii) for each $T \in R^1$, $S_T(N_a \cap D) \subset N_a \cap D$,

(iii) the set mapping PZ is causal on $N_a \cap D$.

Then the admittance A (which is an operator) of $\mathfrak{h}$ is causal on Q(D).

Proof: From Theorem 1.2.4 (a) it follows that, due to (1.6.2) $\mathfrak{h}$ is regular on D. We are going to show that the quasi-inverse $(PZ)^- : [(PZ)(N_a \cap D)]^\circ \to N_a \cap D$ is causal on $[(PZ)(N_a \cap D)]^\circ$. To this end, note first that, by Lemma 1.6.1 and (ii), (iii),

$$S_T(PZ) = S_T(PZ)S_T \qquad (1.6.3)$$

on $N_a \cap D$ for every $T \in R^1$.

Next, choose $T \in R^1$ and $u_1, u_2 \in N_a \cap D$ such that $S_T u_1 \neq S_T u_2$; then $S_T u_j \in N_a \cap D$ by (ii), $j = 1,2$, and by (1.6.2),

$$\langle z_1 - z_2, S_T u_1 - S_T u_2 \rangle \neq 0 \qquad (1.6.4)$$

for any $z_j \in ZS_T u_j$. Since $S_T u_j = PS_T(S_T u_j)$, (1.6.4) yields

$$\langle S_T P z_1 - S_T P z_2, S_T u_1 - S_T u_2 \rangle \neq 0.$$

Hence, for any $u_j \in N_a \cap D$, $S_T u_1 \neq S_T u_2$ and any $p_j \in S_T PZS_T u_j = S_T PZu_j$ we have

$$\langle p_1 - p_2, S_T u_1 - S_T u_2 \rangle \neq 0. \qquad (1.6.5)$$

Now, choose $v_j \in [(PZ)(N_a \cap D)]^\circ$, $j = 1,2$ such that $S_T(PZ)^- v_1 \neq S_T(PZ)^- v_2$ and let $u_j = (PZ)^- v_j$; note that $u_j \in N_a \cap D$. Then $v_j \in PZu_j$, so that $S_T v_j \in S_T PZu_j$. Consequently, we can put $p_j = S_T v_j$ into (1.6.5) and get

$$\langle S_T v_1 - S_T v_2, S_T(PZ)^- v_1 - S_T(PZ)^- v_2 \rangle \neq 0. \qquad (1.6.6)$$

However, (1.6.6) shows that $S_T v_1 - S_T v_2$ cannot be zero; hence, we have the implication $v_j \in [(PZ)(N_a \cap D)]^o$, $j = 1,2$, $S_T(PZ)^- v_1 \neq S_T(PZ)^- v_2 \Rightarrow S_T v_1 \neq S_T v_2$, i.e., the operator $(PZ)^-$ is causal on $[(PZ)(N_a \cap D)]^o$.

To conclude the proof, let $e_1, e_2 \in Q(D)$ be such that $S_T e_1 = S_T e_2$ for some $T \in R^1$. Then $PS_T e_1 = PS_T e_2$, so that by (i), $S_T Pe_1 = S_T Pe_2$; thus, by causality of $(PZ)^-$, $S_T Ae_1 = S_T(PZ)^- Pe_1 = S_T(PZ)^- Pe_2 = S_T Ae_2$. Hence, A is causal on Q(D) what we wanted to show.

Theorem 1.6.2. In Theorem 1.6.1,

(a) assumption (ii) can be replaced by stronger condition

$$S_T D \subset D \tag{1.6.7}$$

for each $T \in R^1$,

(b) assumption (iii) can be replaced by stronger condition "Z is causal on D".

Proof: (a) We have $S_T(N_a \cap D) \subset (S_T N_a) \cap (S_T D) \subset (S_T P\mathcal{H}) \cap D = (PS_T\mathcal{H}) \cap D \subset (P\mathcal{H}) \cap D = N_a \cap D$, i.e., (ii) is satisfied.

(b) If Z is causal on D, it is causal on $N_a \cap D$, too; thus, by Lemma 1.6.1, $S_T Z = S_T Z S_T$ on $N_a \cap D$ for each $T \in R^1$. Hence, by (i), $S_T PZ = PS_T Z = PS_T Z S_T = S_T PZ S_T$, i.e., (iii) holds; hence, the proof.

The requirement (1.6.7) or (ii) in Theorem 1.6.1 may sometimes be inconvenient when dealing with specific cases of networks. Fortunately, this requirement can be replaced by a different assumption; indeed, we have the following:

Theorem 1.6.3. Let $\mathfrak{n} = (Z,a)$ be an a.H.n. over $\mathcal{H}$; assume that

(i) there exist constants $c > 0$ and $p > 1$ such that, for every $T \in R^1$, all $x_j \in N_a \cap D$ and all $z_j \in Zx_j$, $j = 1,2$, we have

$$|\langle z_1 - z_2, S_T(x_1 - x_2)\rangle| \geq c\|S_T(x_1 - x_2)\|^p, \tag{1.6.8}$$

(ii) for each $T \in R^1$, the projection S_T commutes with P.

Then $\mathfrak{n}$ is regular on D and the admittance A of $\mathfrak{n}$ is causal on Q(D).

Proof: First, choose $x_j \in N_a \cap D$ and $z_j \in Zx_j$, $j = 1,2$; since $S_T(x_1 - x_2) \to x_1 - x_2$ as $T \to \infty$, (1.6.8) yields by continuity of inner product,

$$|\langle z_1 - z_2, x_1 - x_2\rangle| \geq c\|x_1 - x_2\|^p.$$

Thus, by Theorem 1.2.4 (b), $\mathfrak{h}$ is regular on D, and in particular, the set-mapping PZ is simple on $N_a \cap D$.

Next, choose a $T \in R^1$ and $x_j \in N_a \cap D$, $j = 1,2$; since by (ii), $S_T(x_1-x_2) = S_TP(x_1-x_2) = PS_T\{S_T(x_1-x_2)\}$, (1.6.8) implies that

$$|\langle w_1-w_2, S_Tx_1-S_Tx_2\rangle| \geq c\|S_Tx_1-S_Tx_2\|^p \qquad (1.6.9)$$

whenever $w_j \in S_TPZx_j$, $j = 1,2$.

Now, let $y_1,y_2 \in [(PZ)(N_a \cap D)]^o$, and put $x_j = (PZ)^- y_j \in N_a \cap D$, $j = 1,2$. Then $y_j \in PZx_j$, and consequently, $S_Ty_j \in S_TPZx_j$. Thus, we can put $w_j = S_Ty_j$, $j = 1,2$ into (1.6.9) and get

$$|\langle S_Ty_1-S_Ty_2, S_T(PZ)^-y_1-S_T(PZ)^-y_2\rangle| \qquad (1.6.10)$$

$$\geq c\|S_T(PZ)^-y_1-S_T(PZ)^-y_2\|^p.$$

Hence, if $S_T(PZ)^-y_1 \neq S_T(PZ)^-y_2$, (1.6.10) shows that $S_Ty_1 \neq S_Ty_2$, i.e., the operator $(PZ)^-$ is causal on $[(PZ)(N_a \cap D)]^o$.

The rest of the proof, i.e., showing that $e_1,e_2 \in Q(D)$, $S_Te_1 = S_Te_2 \Rightarrow S_TAe_1 = S_TAe_2$, is the same as in the proof of Theorem 1.6.1. Hence, A is causal on Q(D).

Remark 1. Note that if the inequality (1.6.8) is satisfied even for all $T \in R^1$, $x_j \in D$ and $z_j \in Zx_j$, $j = 1,2$, the condition (i) is trivially met. Thus, in this case, we have causality of A whatever a may be.

Problems

1.6.1 Let $a < b$, and let φ be a continuous increasing function which maps R^1 onto (a,b). For each $T \in R^1$, define the operator $S_T : L_2[a,b] \to L_2[a,b]$ by

$$(S_Tx)(t) = x(t) \text{ for } a \leq t \leq \varphi(\tau),$$
$$= 0 \quad \text{ for } \varphi(\tau) < t \leq b.$$

Show that $\{S_T : T \in R^1\}$ is a resolution of identity on $L_2[a,b]$.

1.6.2 Let $\mathcal{C}$ be the set of all operators $A : \mathcal{H} \to \mathcal{H}$ which are causal with respect to a given resolution of identity; then

(i) $\mathcal{C}$ is a ring, i.e., $\alpha A + \beta B$, $AB \in \mathcal{C}$ whenever $A,B \in \mathcal{C}$ and α,β are scalars,

(ii) $\mathcal{C}$ is closed in the following sense: If $A_n \in \mathcal{C}$, $n = 1,2,\ldots$ and $A : \mathcal{H} \to \mathcal{H}$ is such that $A_nx \to Ax$ for every $x \in \mathcal{H}$, then $A \in \mathcal{C}$.

1.6.3 Does there exist a linear operator $A : \mathcal{H} \to \mathcal{H}$ such that A is not causal but A^2 is causal?

1.6.4 Let S_T be defined on $L_2[0,\infty)$ as in Problem 1.6.1; give an example of a causal, linear, bounded and 1-to-1 operator A from $L_2[0,\infty)$ onto $L_2[0,\infty)$ such that the inverse A^{-1} is not causal.

1.6.5 Prove the following weaker version of Theorem 1.6.3: Let the abstract Hilbert network $\mathfrak{h} = (Z,a)$ over $\mathcal{H}$ be regular on D; assume that

(i) $\langle z_1 - z_2, S_T(x_1 - x_2)\rangle \neq 0$ for each $T \in R^1$, all $x_j \in N_a \cap D$ such that $S_T x_1 \neq S_T x_2$ and all $z_j \in Zx_j$, $j = 1,2$,

(ii) $S_T P = PS_T$ for each $T \in R^1$.

Then the admittance A of $\mathfrak{h}$ is causal on Q(D).

CHAPTER 2

HILBERT NETWORKS

2.1 Graphs, associated matrices and operators

In this section, we define the concept of a (generally infinite) graph and some matrices and operators associated with such a graph.

First, let us introduce some convenient notation.

Let c_1, c_2 be cardinal numbers such that $c_1, c_2 \leq \aleph_0$; a matrix $M = [m_{ik}]$ (with element m_{ik} standing in the i-th row and k-th column) will be said to be of type $c_1 \times c_2$, if the set of rows (columns) has cardinal c_1, (c_2). A $c \times 1$ matrix ξ, $c \leq \aleph_0$, will be called a c-vector and denoted by $[\xi_k]$ whenever it will be necessary to display its components. The sum of two matrices of the same type will be defined elementwise.

If M and N is of type $c_1 \times c_2$ and $c_2 \times c_3$, respectively, $c_1, c_2, c_3 \leq \aleph_0$, then the product $M \cdot N$ will signify the $c_1 \times c_3$ matrix $[q_{ik}] = \sum_{\mu} m_{i\mu} n_{\mu k}]$ whenever all the elements q_{ik} are defined. More specifically, if $c_2 = \aleph_0$, we require that, for each i,k, the series $\sum_{\mu=1}^{\infty} m_{i\mu} n_{\mu k}$ converges in some sense under consideration.

Note that we will use the dot "$\cdot$" for denoting the matrix multiplication in order to distinguish it from the product of operators defined by such matrices.

Finally, M^T will denote the transpose of M. Observe that we have $(M \cdot N)^T = N^T \cdot M^T$ whenever one of the products is defined.

In the sequel, H will denote a fixed separable Hilbert space. If $c \leq \aleph_0$, we let

$$H^c = \{x : x = [x_k] \text{ is a c-vector}, x_k \in H, \sum_k \|x_k\|^2 < \infty\}, \quad (2.1.1)$$

and

$$\langle x,y \rangle_c = \sum_k \langle x_k, y_k \rangle \quad (2.1.2)$$

for every pair $x,y \in H^c$. It is easy to see that H^c with the inner product $\langle \cdot,\cdot \rangle_c$ is again a separable Hilbert space.

Moreover, a c-vector $x \in H^c$ will be called simple, if all but finitely many of its components are zero.

In particular, if $H = R^1$ (real line or complex plane), then H^c will become the space ℓ_2 provided $c = \aleph_0$, and the n-dimensional Euclidean space R^n whenever $c = n < \aleph_0$. Note that, using the vector notation introduced above, we have for any $\xi, \eta \in R^c$,

$$\langle \xi, \eta \rangle_c = \sum_k \xi_k \bar{\eta}_k = \xi^T \cdot \bar{\eta}, \text{ and } \|\xi\|_c^2 = \xi^T \cdot \bar{\xi}.$$

Let $\mathcal{B}$ and $\mathcal{V}$ be nonempty sets having cardinals c_2 and c_1, respectively, $c_1, c_2 \leq \aleph_0$. Since $\mathcal{B}$ and $\mathcal{V}$ are at most countable we will assume that all elements in $\mathcal{B}$ are ordered into a fixed (finite or infinite) sequence (b_i), and elements in $\mathcal{V}$ into a fixed sequence (v_k).

A mapping $G : \mathcal{B} \to \mathcal{V} \times \mathcal{V}$ will be called an oriented graph, if

(i) for every $v_k \in \mathcal{V}$ at least one of the sets

$$\mathcal{B}_k^+ = \{b_j : b_j \in \mathcal{B}, \ Gb_j = (v_\alpha, v_k)\}$$

and

$$\mathcal{B}_k^- = \{b_j : b_j \in \mathcal{B}, \ Gb_j = (v_k, v_\alpha)\}$$

is nonempty,

(ii) $\{b_i : b_i \in \mathcal{B}, \ Gb_i = (v_\alpha, v_\alpha) \text{ for any } v_\alpha \in \mathcal{V}\} = \emptyset$.

Moreover, an oriented graph G will be called locally finite, if

(iii) for each $v_k \in \mathcal{V}$, the sets $\mathcal{B}_k^+$ and $\mathcal{B}_k^-$ are finite.

An oriented graph G will be called finite, if $\mathcal{B}$ and $\mathcal{V}$ are finite sets. (Clearly, G finite $\Rightarrow$ G locally finite.)

Also, $\mathcal{B}$ will be called the set of branches of G, and $\mathcal{V}$ the set of vertices (nodes).

From our definition it follows that, due to (i), G does not contain "isolated nodes", and, due to (ii), branches "beginning and ending" at the same node. Also, if G is locally finite, then each node $v_k \in \mathcal{V}$ is incident with only a finite number of branches (witness (iii)).

Observe that for any oriented graph G we have $c_1 \leq 2c_2$.

Let G be an oriented graph and assume that, as mentioned earlier, the branches of G are enumerated in a fixed manner, and the same is true for nodes; define the $c_2 \times c_1$ matrix $d = [d_{ik}]$ (called the incidence matrix of G) by

$$d_{ik} = 1, \text{ if } Gb_i = (v_\alpha, v_k) \tag{2.1.3}$$
$$= -1, \text{ if } Gb_i = (v_k, v_\alpha)$$
$$= 0, \text{ if } Gb_i = (v_\alpha, v_\beta) \text{ with } \alpha \neq k,\ \beta \neq k.$$

It is clear that, due to the assumptions (i), (ii), the definition of d is meaningful, i.e., for any pair of indices i,k, exactly one relation $Gb_i = \ldots$ in (2.1.3) is satisfied.

Note also the fact that each row of d contains exactly one element 1, one element -1 and all other elements are zero. Moreover, if G is locally finite, then, since each node v_k is incident with finitely many branches b_i by (iii), each column of d contains finitely many nonzero elements.

Let G be a locally finite oriented graph with incidence matrix d; choose an arbitrary diagonal $c_1 \times c_1$ matrix K, $K = \mathrm{diag}(k_1, k_2, k_3, \ldots)$ such that $k_i \neq 0$ for every i and such that the $c_1 \times c_2$ matrix

$$a = K \cdot d^T = [a_{ik}] \tag{2.1.4}$$

has the property that $\sum_{ik} |a_{ik}|^2 < \infty$. Clearly, such a choice of K is always possible, since each row of d^T contains only finitely many nonzero elements due to the local finiteness of G.

Now, define the mapping $\tilde{a} : R^{c_2} \to R^{c_1}$ by

$$\tilde{a}\xi = a \cdot \xi, \tag{2.1.5}$$

and let $N_{\tilde{a}} = \{\xi : \xi \in R^{c_2}, \tilde{a}\xi = 0\}$.

Note that, due to (2.1.4), $\tilde{a}$ truly maps R^{c_2} into R^{c_1}, and $\tilde{a} \in [R^{c_2}, R^{c_1}]$.

Indeed, let $\xi = [\xi_k] \in R^{c_2}$; then, for any index i, the element q_i of $a \cdot \xi$ equals $\sum_k a_{ik}\xi_k$. If $c_2 = \aleph_0$, then putting $s_n^i = \sum_{k=1}^{n} a_{ik}\xi_k$, we have for $m > n$,

$$|s_m^i - s_n^i|^2 = \Big|\sum_{k=n+1}^{m} a_{ik}\xi_k\Big|^2 \le \Big(\sum_{k=m+1}^{m} |a_{ik}|^2\Big)\Big(\sum_{k=n+1}^{m} |\xi_k|^2\Big) \le \Big(\sum_{ik} |a_{ik}|^2\Big)\Big(\sum_{k=n+1}^{m} |\xi_k|^2\Big);$$

hence, the series $\sum_{k=1}^{\infty} a_{ik}\xi_k$ converges for every i, i.e., $a \cdot \xi$ is well-defined.

Moreover, we have

$$\sum_i |q_i|^2 = \sum_i \Big|\sum_k a_{ik}\xi_k\Big|^2 \le \sum_i \Big(\sum_k |a_{ik}|^2\Big)\Big(\sum_k |\xi_i|^2\Big) = \sum_{ik} |a_{ik}|^2 \cdot \|\xi\|^2;$$ consequently,

$[q_i] = a \cdot \xi \in R^{c_1}$ and $\|a \cdot \xi\| \le \lambda \|\xi\|$ with $\lambda^2 = \sum_{ik} |a_{ik}|^2$. Thus, $\tilde{a} \in [R^{c_2}, R^{c_1}]$.

Due to continuity of $\tilde{a}$, $N_{\tilde{a}}$ is a closed subspace of R^{c_2}; it is easy to see that $N_{\tilde{a}}$ does not depend on the choice of matrix K, since

$$\tilde{a}\xi = 0 \Leftrightarrow d^T \cdot \xi = 0 \qquad (2.1.6)$$

whenever ξ is in R^{c_2}.

In the sequel we will assume that $N_{\tilde{a}} \neq \{0\}$, (i.e., G contains at least one loop); the meaning of this assumption will be clarified later.

Since $N_{\tilde{a}}$ is a closed nontrivial subspace of R^{c_2}, it is a separable Hilbert space of its own right. Thus, choose a fixed orthonormal basis $\{\xi^i\}$ in $N_{\tilde{a}}$, and let $c_o \leq \aleph_o$ be the cardinal of $\{\xi^i\}$. Moreover, let X be the $c_2 \times c_o$ matrix, whose columns are all the c_2-vectors ξ^i ordered in some fixed fashion.

It is clear that

$$a \cdot X = 0 \qquad (2.1.7)$$

Define now the mapping $\tilde{X} : R^{c_o} \to R^{c_2}$ by

$$\tilde{X}\eta = X \cdot \eta. \qquad (2.1.8)$$

Then we have

Lemma 2.1.1.

(a) $\tilde{X} \in [R^{c_o}, R^{c_2}]$ and $N_{\tilde{a}} = \tilde{X}R^{c_o}$.

(b) $\|\tilde{X}\xi\| = \|\xi\|$ for each $\xi \in R^{c_o}$.

(c) If $\tilde{X}^* \in [R^{c_2}, R^{c_1}]$ is the adjoint of $\tilde{X}$, then

$$\tilde{X}^*\tilde{X} = I, \qquad (2.1.9)$$

where I is the identity operator on R^{c_o}, and $\tilde{X}^*\xi = \bar{X}^T \cdot \xi$ for every $\xi \in R^{c_2}$.

Proof: If $\eta = [\eta_i] \in R^{c_o}$, then (2.1.8) means that $\tilde{X}\eta = \sum_i \eta_i \xi^i$, where the c_2-vectors ξ^i constitute an orthonormal basis in $N_{\tilde{a}}$; hence, by Riesz-Fischer theorem [1], $N_{\tilde{a}} = \tilde{X}R^{c_o}$. In view of this fact, (b) is nothing else than the Parseval's equality for the basis $\{\xi^i\}$; also, (b) proves the boundedness of $\tilde{X}$. Finally, (c) is a standard result. (After all, (c) is a special case of proposition (v) in the next Lemma).

Define now the operator $\hat{a} : H^{c_2} \to H^{c_1}$ by

$$\hat{a}x = a\cdot[x_k], \tag{2.1.10}$$

where $x = [x_k]$ is any c_2-vector in H^{c_2}.

Using the same reasoning as for operator $\tilde{a}$, we see readily that $\hat{a} \in [H^{c_2}, H^{c_1}]$.

Furthermore, let

$$N_{\hat{a}} = \{x : x \in H^{c_2}, \hat{a}x = 0\}. \tag{2.1.11}$$

Again, since $\hat{a}$ is bounded, $N_{\hat{a}}$ is a closed subspace of H^{c_2}. Moreover, $N_{\hat{a}}$ does not depend on the choice of K, because we clearly have

$$\hat{a}x = 0 \Leftrightarrow d^T\cdot x = 0 \tag{2.1.12}$$

whenever $x \in H^{c_2}$.

Now we are ready to prove the following crucial result:

<u>Lemma 2.1.2</u>. Let the operator $\hat{X}$ be defined on H^{c_o} by

$$\hat{X}z = X\cdot[z_k], \ z = [z_k]; \tag{2.1.13}$$

then we have:

(i) $\hat{X} \in [H^{c_o}, H^{c_2}]$.

(ii) $N_{\hat{a}} = \hat{X}H^{c_o}$ and $\hat{X}$ is 1-to-1.

(iii) $\hat{a}\hat{X} = 0$ on H^{c_o}.

(iv) $\|\hat{X}z\|_{c_2} = \|z\|_{c_o}$ for every $z \in H^{c_o}$.

(v) If $\hat{X}^* \in [H^{c_2}, H^{c_o}]$ is the adjoint of $\hat{X}$, then

$$\hat{X}^*\hat{X} = I, \tag{2.1.14}$$

where I is the identity operator on H^{c_o}, and

$$\hat{X}^*v = \bar{X}^T\cdot v \tag{2.1.15}$$

for all $v \in H^{c_2}$.

(vi) $N_{\hat{X}^*} = N_{\hat{a}}^{\perp}$, where $N_{\hat{X}^*} = \{x : x \in H^{c_2}, \hat{X}^*x = 0\}$.

(vii) If P is the orthogonal projection from H^{c_2} onto $N_{\hat{a}}$, then $P = \hat{X}\hat{X}^*$.

<u>Proof</u>: First let us show that the definition (2.1.13) is meaningful. As before, let ξ^k be the columns of X. Choose arbitrarily $z = [z_k] \in H^{c_o}$; by definition of H^{c_o},

$$\sum_k \|z_k\|^2 = \|z\|^2_{c_o} < \infty. \tag{2.1.16}$$

Let $z_{k_1}, z_{k_2}, z_{k_3}, \ldots$ be the subsequence of $z_1, z_2, z_3, \ldots$, which consists of all nonzero z_k's. Then, by (2.1.13),

$$\hat{X}z = \sum_j \eta^{k_j} \|z_{k_j}\|, \tag{2.1.17}$$

where

$$\eta^{k_j} = \xi^{k_j} \frac{z_{k_j}}{\|z_{k_j}\|} = [\xi_{ik_j} \frac{z_{k_j}}{\|z_{k_j}\|}],$$

and where we use the notation $\xi^k \doteq [\xi_{ik}]$. However, $\{\eta^{k_j} : j = 1,2,\ldots\}$ is an orthonormal family in H^{c_2}. Indeed,

$$\langle \eta^{k_j}, \eta^{k_{j'}} \rangle_{c_2} = \sum_i \langle \xi_{ik_j} \frac{z_{k_j}}{\|z_{k_j}\|}, \xi_{ik_{j'}} \frac{z_{k_{j'}}}{\|z_{k_{j'}}\|} \rangle$$

$$= \frac{\langle z_{k_j}, z_{k_{j'}} \rangle}{\|z_{k_j}\| \|z_{k_{j'}}\|} \sum_i \xi_{ik_j} \bar{\xi}_{ik_{j'}} = \frac{\langle z_{k_j}, z_{k_{j'}} \rangle}{\|z_{k_j}\| \|z_{k_{j'}}\|} \langle \xi^{k_j}, \xi^{k_{j'}} \rangle.$$

Hence, $\langle \eta^{k_j}, \eta^{k_{j'}} \rangle_{c_2} = 0$ for $j \neq j'$, and $\langle \eta^{k_j}, \eta^{k_j} \rangle_{c_2} = 1$. Since

$\sum_j \|z_{k_j}\|^2 = \sum_k \|z_k\|^2 = \|z\|^2_{c_o} < \infty$ by (2.1.16), it follows that $\hat{X}z \in H^{c_2}$, and, by

Parseval's equality, $\|\hat{X}z\|^2_{c_2} = \sum_j \|z_{k_j}\|^2 = \|z\|^2_{c_o}$. Hence, $\hat{X} \in [H^{c_o}, H^{c_2}]$, and

$\|\hat{X}z\|_{c_2} = \|z\|_{c_o}$ for each $z \in H^{c_o}$, i.e., (i) and (iv) hold.

Next, let $h \in H$ and let the c_o-vector z_o be defined by $z_o = [0,0,\ldots,0,h,0,\ldots]^T$, where the element h stands at the k-th place; then $z_o \in H^{c_o}$ and we have $\hat{a}\hat{X}z_o = a\cdot(X\cdot z_o) = a\cdot(\xi^k h) = 0$. Thus, if $z' \in H^{c_o}$ is simple, we have by linearity of $\hat{a}$ and $\hat{X}$ that $\hat{a}\hat{X}z' = 0$. Now, if $z = [z_k] \in H^{c_o}$, there exists a sequence of simple vectors $z^n \in H^{c_o}$ such that $z^n \to z$. (We can put, for example, $z^n = [z_1, z_2, \ldots, z_n, 0, 0, \ldots]^T$). Since $\hat{a}$ is continuous and we have already proved the continuity of $\hat{X}$, it follows that $0 = \hat{a}\hat{X}z_n \to \hat{a}\hat{X}z$; hence, $\hat{a}\hat{X}z = 0$, i.e., $\hat{a}\hat{X} = 0$ on H^{c_o} and (iii) is proven.

Let us now prove (ii). First of all (iv) shows that $\hat{X}$ is 1-to-1. To prove that the range of $\hat{X}$ is $N_{\hat{a}}$, choose some $x = [x_i] \in N_{\hat{a}}$; then $\hat{a}x = a\cdot[x_i] = 0$. Let $\{f_j\}$ be some fixed orthonormal basis in H; since H is separable, $\{f_j\}$ is at most countable. Since each $x_i \in H$, there exist numbers s_{ik} such that

$$x_i = \sum_k s_{ik} f_k, \tag{2.1.18}$$

the series being convergent in H, and $\|x_i\|^2 = \sum_k |s_{ik}|^2$. Thus, we have

$$\sum_i \sum_k |s_{ik}|^2 = \sum_i \|x_i\|^2 = \|x\|^2_{c_2} < \infty. \qquad (2.1.19)$$

Define the c_2-vector s^k by $s^k = [s_{1k}, s_{2k}, s_{3k}, \ldots]^T$; then $\|s^k\|^2 = \sum_i |s_{ik}|^2 < \infty$ by (2.1.19), i.e., $s^k \in R^{c_2}$.

Now we have

$$x = \sum_k (s^k f_k), \qquad (2.1.20)$$

where the series converges in H^{c_2}. Indeed, we clearly have $s^k f_k \in H^{c_2}$. Moreover, denoting $S_n = \sum_{k=1}^{n} (s^k f_k)$, then for $m > n$ we have by orthogonality of the f_j's,

$$\|S_m - S_n\|^2 = \| \sum_{k=n+1}^{m} (s^k f_k) \|^2 = \sum_{j,k=n+1}^{m} \langle s^k f_k, s^j f_j \rangle_{c_2}$$

$$= \sum_{j,k=n+1}^{m} (\sum_i \langle s_{ik} f_k, s_{ij} f_j \rangle)$$

$$= \sum_{j,k=n+1}^{m} (\sum_i s_{ik} \bar{s}_{ij} \langle f_k, f_j \rangle)$$

$$= \sum_{k=n+1}^{m} (\sum_i |s_{ik}|^2);$$

this shows by (2.1.19), however, that S_n is Cauchy in H^{c_2}, i.e., there exists $y \in H^{c_2}$ such that $S_n \to y = \sum_k (s^k f_k)$. On the other hand, due to (2.1.18) it follows that necessarily $y = x$, and (2.1.20) is confirmed.

Now, by continuity of $\hat{a}$ it follows from (2.1.20) that $\hat{a}x = 0 = \sum_k \hat{a}(s^k f_k)$. Since clearly $\hat{a}(s^k f_k) = (a \cdot s^k) f_k$, we have

$$\sum_k (a \cdot s^k) f_k = 0, \qquad (2.1.21)$$

where the series converges in H^{c_2}. Denoting $a \cdot s^k = [m_i^k]$, (2.1.21) implies that $\sum_k m_i^k f_k = 0$ in H for $i = 1,2,\ldots$; consequently, since $\{f_j\}$ is a basis in H, $m_i^k = 0$ for $k = 1,2,\ldots$ and $i = 1,2,\ldots$. Hence, $a \cdot s^k = 0$ for each k, and $s^k \in R^{c_2}$.

By Lemma 2.1.1 (a), however, for each k there exists vector $\eta^k = [\eta_{ik}] \in R^{c_o}$ such that

$$s^k = X \cdot \eta^k, \qquad (2.1.22)$$

and, by (b),

$$\|s^k\| = \|\eta^k\|. \qquad (2.1.23)$$

Hence, we have by (2.1.23), $\sum_i |s_{ik}|^2 = \sum_i |\eta_{ik}|^2$, and consequently, by (2.1.19),

$$\sum_k \sum_i |\eta_{ik}|^2 = \sum_k \sum_i |s_{ik}|^2 = \|x\|^2_{c_2} < \infty. \tag{2.1.24}$$

Define now the element $u \in H^{c_o}$ by

$$u = \sum_k (\eta^k f_k). \tag{2.1.25}$$

This definition is meaningful, since the series in (2.1.25) converges in H^{c_o}. Indeed, due to (2.1.24) we can easily show as above that the partial sums $S'_n = \sum_{k=1}^{n} (\eta^k f_k)$ constitute a Cauchy sequence in H^{c_o}.

Next, since $\hat{X}$ is continuous, (2.1.25) yields by (2.1.22) and (2.1.20),

$$\hat{X}u = \sum_k \hat{X}(\eta^k f_k) = \sum_k X\cdot(\eta^k f_k) = \sum_k (X\cdot\eta^k) f_k = \sum_k (s^k f_k) = x.$$

Hence, $x \in \hat{X}H^{c_o}$, i.e., we have proved that $N_{\hat{a}} \subset \hat{X}H^{c_o}$.

To prove the converse inclusion, choose some $x \in \hat{X}H^{c_o}$; then $x = \hat{X}w$ for some $w \in H^{c_o}$, and consequently, by (iii), $\hat{a}x = \hat{a}\hat{X}w = 0$, i.e., $x \in N_{\hat{a}}$; hence $\hat{X}H^{c_o} \subset N_{\hat{a}}$ and the proposition (ii) is proven.

As for the proof of (v), choose arbitrarily $h,h' \in H$ and define the vector $x_o \in H^{c_o}$ by $x_o = [0,0,\ldots,0,h,0,\ldots]^T$ with h standing at the k-th place, and vector $u_o \in H^{c_2}$ by $u_o = [0,0,\ldots,h',0,\ldots]^T$, where h' stands at the j-th place. Then we have

$$\langle \hat{X}x_o, u_o \rangle_{c_2} = \langle [\xi_{ik}h], u_o \rangle_{c_2} = \langle \xi_{jk}h, h' \rangle$$

$$= \langle h, \bar{\xi}_{jk}h' \rangle = \langle x_o, \bar{X}^T \cdot u_o \rangle_{c_o}.$$

On the other hand, by definition of the adjoint operator, $\langle \hat{X}x_o, u_o \rangle_{c_2} = \langle x_o, \hat{X}^* u_o \rangle_{c_o}$; consequently

$$\langle x_o, \hat{X}^* u_o - \bar{X}^T \cdot u_o \rangle_{c_o} = 0. \tag{2.1.26}$$

By linearity it follows that (2.1.26) holds for any simple x_o, u_o.

Next, if $x \in H^{c_o}$, there exist simple vectors $x^n \in H^{c_o}$ such that $x^n \to x$, and thus, by continuity of inner product and (2.1.26),

$$\langle x, \hat{X}^* u_o - \bar{X}^T \cdot u_o \rangle_{c_o} = 0. \tag{2.1.27}$$

Hence, if u_o is simple and we set $x = \hat{X}^* u_o - \bar{X}^T \cdot u_o$, (2.1.27) yields

$$\hat{X}^* u_o - \bar{X}^T \cdot u_o = 0. \tag{2.1.28}$$

Now, let $u = [u_k] \in H^{c_2}$, and put $u^n = [u_1, u_2, \ldots, u_n, 0, 0, \ldots]^T$; then each $u^n \in H^{c_2}$ is simple and $u^n \to u$ in H^{c_2}. Since $\hat{X}^*$ is continuous, (2.1.28) yields

$$\bar{X}^T \cdot u^n = \hat{X}^* u^n \to \hat{X}^* u. \tag{2.1.29}$$

This, however, means that the product $\bar{X}^T \cdot u$ is meaningful and equals $\hat{X}^* u$. (In more detail, letting $\bar{X}^T = [\xi_{ik}]$, then the i-th element q_i^n of $\bar{X}^T \cdot u^n$ is $\sum_{k=1}^{n} \xi_{ik} u_k$. Since convergence in H^{c_2} implies the "elementwise" convergence, (2.1.29) means that the sequence q_i^n converges in H, i.e., $\sum_{k=1}^{\infty} \xi_{ik} u_k$ converges in H. This, however, is exactly what is required in the definition of a product of infinite matrices.)

To prove (2.1.14), choose $h \in H$ and define the c_o-vector x_o by $x_o = [0,0,\ldots,0,h,0,\ldots]^T$, where h stands at the k-th place; then $x_o \in H^{c_o}$ and we have $\hat{X}^* \hat{X} x_o = \bar{X}^T \cdot (X \cdot x_o) = \bar{X}^T \cdot (\xi^k h) = x_o$. Thus, by linearity, $\hat{X}^* \hat{X} x' = x'$ for any simple $x' \in H^{c_o}$. Now, if $x \in H^{c_o}$, then there exists a simple $x^n \in H^{c_o}$ such that $x^n \to x$; consequently, by continuity of $\hat{X}^*$ and $\hat{X}$, $\hat{X}^* \hat{X} x^n = x^n \to \hat{X}^* \hat{X} x$. Hence, $\hat{X}^* \hat{X} x = x$, i.e., $\hat{X}^* \hat{X} = I$ on H^{c_o}, and (v) is completely proven.

Referring to (vi), we have by Lemma 1.4.2 and (ii), $N_{\hat{X}^*} = (\hat{X} H^{c_o})^{\perp} = N_{\hat{a}}^{\perp}$.

Finally, let us prove (vii). To this end, observe first that $\hat{X}^*$ maps H^{c_2} onto H^{c_o}. This follows readily from (2.1.14), which means that $\hat{X}$ is a right inverse of $\hat{X}$.

Next, let $Q : H^{c_2} \to H^{c_2}$ be defined by $Q = \hat{X}\hat{X}^*$; then we have by (2.1.14), $Q^2 = \hat{X}\hat{X}^*\hat{X}\hat{X}^* = \hat{X}\hat{X}^* = Q$. Also, $Q^* = (\hat{X}\hat{X}^*)^* = \hat{X}\hat{X}^* = Q$. Hence, Q is an orthogonal projection.

Now, we are going to show that $QH^{c_2} = N_{\hat{a}}$. Indeed, if $x \in QH^{c_2}$, then $x = \hat{X}\hat{X}^* y$ for some $y \in H^{c_2}$. By (iii), however, $\hat{a}x = \hat{a}\hat{X}(\hat{X}^* y) = 0$, i.e., $x \in N_{\hat{a}}$; hence, $QH^{c_2} \subset N_{\hat{a}}$. Conversely, let $x \in N_{\hat{a}}$; then $x = \hat{X} y$ for some $y \in H^{c_o}$ according to (ii). However, since $\hat{X}^*$ is onto H^{c_o}, there exist some $z \in H^{c_2}$ such that $y = \hat{X}^* z$; consequently, $x = \hat{X}\hat{X}^* z = Qx$, i.e., $N_{\hat{a}} \subset QH^{c_2}$. Hence, Q is the orthogonal projection of H^{c_2} onto $N_{\hat{a}}$, and the proof of Lemma 2.1.2 is completed.

Problems

2.1.1 Referring to (2.1.1) and (2.1.2), prove in detail that H^c, $c \leq \aleph_o$, is a separable Hilbert space whenever H is.

2.1.2 Let G be a locally finite oriented graph, let $\tilde{a}$ have the usual meaning and assume that $N_{\tilde{a}} \neq \{0\}$. Let X_j, j = 1,2 be a $c_2 \times c_o$ matrix whose columns constitute an orthonormal basis in $N_{\tilde{a}}$, and let $\hat{X}_j : H^{c_o} \to H^{c_2}$ be defined by (2.1.13). Then there exists a linear, bounded, 1-to-1 operator S from H^{c_o}

onto H^{c_o} with $||S|| = 1$ such that $\hat{X}_2 = \hat{X}_1 S$.

2.1.3 Let G be a finite oriented graph, and let $N_{\tilde{a}} \neq \{0\}$. Let Y be a $c_2 \times c_o$ matrix, whose columns constitute a (not necessarily orthonormal) basis in $N_{\tilde{a}}$. Define $\hat{Y} : H^{c_o} \to H^{c_2}$ by $\hat{Y}z = Y \cdot z$. Then

(i) $\hat{Y} \in [H^{c_o}, H^{c_2}]$,

(ii) $N_{\hat{a}} = \hat{Y}H^{c_o}$ and $\hat{Y}$ is 1-to-1,

(iii) $\hat{a}\hat{Y} = 0$ on H^{c_o},

(iv) $N_{\hat{Y}*} = N_{\hat{a}}^{\perp}$, where $N_{\hat{Y}*} = \{u : u \in H^{c_2}, \hat{Y}*u = 0\}$.

2.2 Solution of a Hilbert network

In this section we will define the concept of a Hilbert network, the concept of a solution and we will establish several theorems on the existence and uniqueness of a solution.

Let H be a fixed Hilbert space, and let G be an oriented locally finite graph having the set of branches $\mathfrak{B}$ and the set of vertices $\mathfrak{V}$; let $\mathfrak{B}$ and $\mathfrak{V}$ have the cardinal $c_2 \leq \aleph_o$ and $c_1 \leq \aleph_o$, respectively, and let operators $\hat{a} \in [H^{c_2}, H^{c_1}]$ and $\hat{X} \in [H^{c_o}, H^{c_2}]$ be defined as in Section 2.1.

Furthermore, let $D \subset H^{c_2}$, $D \neq \phi$, and let $\hat{Z} : D \to \mathfrak{S}(H^{c_2})$ be a set mapping. Then the ordered pair $\hat{n} = (\hat{Z}, G)$ will be called a Hilbert network.

A network $\hat{n}$ will be called infinite (finite), if G is an infinite (finite) graph, i.e., if $c_2 = \aleph_o$ (or $c_2 < \aleph_o$).

If a vector $x = [x_k] \in H^{c_2}$ has some relevance to the network $\hat{n}$, it will be understood that the k-th component x_k of x is associated with the branch $b_k \in \mathfrak{B}$, where the branches in $\mathfrak{B}$ have been enumerated in some fixed manner as indicated in Section 2.1. Thus, using physical terminology, if $e = [e_k] \in H^{c_2}$ has the meaning of a vector of voltages in $\hat{n}$, then $e_k \in H$ will be interpreted as the voltage source in the branch b_k; similarly, if $i = [i_k] \in H^{c_2}$ is the vector of currents in $\hat{n}$, then the element $i_k \in H$ will have the meaning of a current in the branch b_k. Also, if we put $u = [u_k] \in \hat{Z}i \in H^{c_2}$, then the k-th component $u_k \in H$ will signify the voltage drop in b_k caused by currents $i_1, i_2, i_3, \ldots$.

A special case of a Hilbert network is obtained if $\hat{Z}$ is an operator described by a matrix. Indeed, assume that to each ordered pair $(b_i, b_k) \in \mathfrak{B} \times \mathfrak{B}$ there is assigned an operator $Z_{ik} : H \to H$ such that the mapping $\hat{Z}$ defined on H^{c_2} by

$$\hat{Z}x = [Z_{ik}] \cdot [x_k] \qquad (2.2.1)$$

maps H^{c_2} into itself. In this case, the voltage drop u_ℓ in the branch b_ℓ is additive in drops caused by currents $i_1, i_2, \ldots$, i.e., $u_\ell = \sum_k Z_{\ell k} i_k$. Of course, if each operator Z_{ik} is defined on H and is linear, then $\hat{Z}$ is linear on H^{c_2} and vice versa.

Definition 1. Let $\hat{\mathfrak{n}} = (\hat{Z}, G)$ be a Hilbert network, and let $e \in H^{c_2}$; an element $i \in H^{c_2}$ will be called a solution of $\hat{\mathfrak{n}}$ corresponding to e, if i is a solution of the abstract Hilbert network $\hat{\mathfrak{n}} = (\hat{Z}, \hat{a})$ over H^{c_2} corresponding to e, i.e., if

K_1 : there exists a $v \in \hat{Z}i$ such that $v - e \in N_{\hat{a}}^{\perp}$,

K_2 : $i \in N_{\hat{a}} \cap D$.

The network $\hat{\mathfrak{n}}$ will be called regular on D, if $\mathfrak{n}$ is regular on D. The admittance A of $\mathfrak{n}$, $A : Q(D) \to \mathfrak{S}(N_{\hat{a}} \cap D)$, will be called the admittance of $\hat{\mathfrak{n}}$.

Obviously, we have defined the solution of a Hilbert network $\hat{\mathfrak{n}}$ as a solution of the associated abstract network $\mathfrak{n}$. The question is, however, what justifies such a definition. Having the classical Kirchhoff laws in mind, it would be perhaps more natural to define the solution concept for $\hat{\mathfrak{n}}$ by the following definition:

Definition 2. Let $\hat{\mathfrak{n}} = (\hat{Z}, G)$ be a Hilbert network, and let $e \in H^{c_2}$; an element $i \in H^{c_2}$ will be called a classical solution of $\hat{\mathfrak{n}}$ corresponding to e, if

K_1^+ : there exists an element $v \in \hat{Z}i$ such that

$$\bar{\gamma}^T \cdot (v - e) = 0 \tag{2.2.2}$$

for every $\gamma \in R^{c_2}$ satisfying the equation $a \cdot \gamma = 0$,

K_2^+ : $i \in D$ and $a \cdot i = 0$.

Observe that this definition is meaningful. Indeed, if $e, i \in H^{c_2}$ and $v \in \hat{Z}i \subset H^{c_2}$, then $v - e = u$ is also in H^{c_2}, and consequently, $\bar{\gamma}^T \cdot u$ is defined and belongs to H for any $\gamma \in R^{c_2}$. (If $u = [u_k]$, $\gamma = [\gamma_k]$, then elements s_n defined by $s_n = \sum_{k=1}^{n} \bar{\gamma}_k u_k$ constitute a Cauchy sequence in H by Schwartz inequality, and thus, $\bar{\gamma}^T \cdot u = \sum_k \bar{\gamma}_k u_k$ is well-defined.) Also, it is easy to see (cf. Section 2.1, (2.1.6)) that the classical solution does not depend on the choice of the matrix K.

Fortunately, it turns out that both solution concepts coincide; to prove this fact, we will need the following lemma:

Lemma 2.2.1. Let $\hat{\mathfrak{n}}$ be a Hilbert network and let $e \in H^{c_2}$; then $i \in H^{c_2}$ is a classical solution of $\hat{\mathfrak{n}}$ corresponding to e, if and only if

(i) there exists $v \in \hat{Z}i$ such that $v-e \in N_{\hat{X}*}$,

(ii) $i \in N_{\hat{a}} \cap D$.

Proof: (1) Let i be a classical solution of $\hat{\mathfrak{n}}$ corresponding to $e \in H^{c_2}$; then by K_1^+, there exists $v \in \hat{Z}i$ such that $\bar{\gamma}^T \cdot (v-e) = 0$ for every $\gamma \in N_{\tilde{a}}$. (See (2.1.5)). Since each column ξ^j of X is in $N_{\tilde{a}}$, we have $(\bar{\xi}^j)^T \cdot (v-e) = 0$ for every j, i.e., $\bar{X}^T \cdot (v-e) = 0$. Thus, by Lemma 2.1.2 (v), $v-e \in N_{\hat{X}*}$ and (i) is satisfied. The condition K_2^+ is clearly the same as (ii).

(2) Conversely, let (i) and (ii) hold. Then by Lemma 2.2 (v) we have

$$\bar{X}^T \cdot u = 0 \tag{2.2.3}$$

with $u = v-e \in H^{c_2}$. Now, let $\gamma \in N_{\tilde{a}}$, i.e., $\gamma \in R^{c_2}$ and $a \cdot \gamma = 0$; we are going to show that $\bar{\gamma}^T \cdot u = 0$.

To do this, observe first that by Lemma 2.2.1 (a) there exists $\eta \in R^{c_o}$ such that $\gamma = X \cdot \eta$. Let $\eta_b = [0,0,\ldots,0,1,0,\ldots]^T$ where 1 stands at the k-th place, and η_b is a c_o-vector; clearly $\eta_b \in R^{c_o}$. Then by (2.2.3), we have $\bar{\eta}_b^T \cdot (\bar{X}^T \cdot u) = 0$, i.e., $(u^T \cdot \bar{X}) \cdot \bar{\eta}_b = 0$. However, $0 = (u^T \cdot \bar{X}) \cdot \bar{\eta}_b = [\sum_i u_i \bar{\xi}_{i1}, \sum_i u_i \bar{\xi}_{i2}, \ldots] \cdot \bar{\eta}_b = \sum_i u_i \bar{\xi}_{ik}$, and $u^T \cdot (\bar{X} \cdot \bar{\eta}_b) = \sum_i u_i \bar{\xi}_{ik}$. Hence,

$$u^T \cdot (\bar{X} \cdot \bar{\eta}_b) = 0. \tag{2.2.4}$$

Thus, it follows by linearity that (2.2.4) is true for every simple $\eta_b \in R^{c_o}$.

On the other hand, define $U : R^{c_2} \to H$ by

$$U\lambda = u^T \cdot \lambda = \sum_i \lambda_i u_i; \tag{2.2.5}$$

according to the Schwartz inequality, this definition is meaningful, and we have

$$\|U\lambda\| \le \sum_i |\lambda_i| \, \|u_i\| \le \sqrt{\sum_i |\lambda_i|^2} \cdot \sqrt{\sum_i \|u_i\|^2} = \|\lambda\| \cdot \|u\|_{c_2},$$

i.e., $U \in [R^{c_2}, H]$.

Now for the vector $\eta \in R^{c_o}$ above, there exist simple vectors $\eta_n \in R^{c_o}$ such that $\eta_n \to \eta$ in R^{c_o}. Then, by continuity of $\tilde{X}$,

$$X \cdot \eta = \tilde{X}\eta_n \to \tilde{X}\eta = X \cdot \eta \text{ in } R^{c_2},$$

so that $\bar{X} \cdot \bar{\eta}_n \to \bar{X} \cdot \bar{\eta}$. Thus, by continuity of U and (2.2.4),

$$0 = u^T\cdot(\bar{X}^T\cdot\bar{\eta}_n) = U(\bar{X}^T\cdot\bar{\eta}^n) \to U(\bar{X}\cdot\bar{\eta}) = u^T\cdot(\bar{X}\cdot\bar{\eta}),$$

i.e., $u^T\cdot\bar{\gamma} = 0$. Hence, K_1^+ holds, and since (ii) is the same as K_2^+, our Lemma is proven.

From Lemma 2.2.1 we get immediately the following important result:

Theorem 2.2.1. For any Hilbert network $\hat{n}$, the solution and classical solution coincide.

Proof: Since $N_{\hat{X}*} = N_{\hat{a}}^{\perp}$ by Lemma 2.1.2 (vi), we have by Definitions 1,2 and Lemma 2.2.1 for any $i \in N_{\hat{a}} \cap D$, K_1 holds $\Leftrightarrow$ K_1^+ holds; hence, the proof.

Remark 1. We have proved Theorem 2.2.1 under the assumption that $N_{\tilde{a}} \neq \{0\}$, which has been made earlier, after we have defined the operator $\tilde{a}$ and the subspace $N_{\tilde{a}}$. Clearly, if $N_{\tilde{a}} \neq \{0\}$, then $N_{\tilde{a}}$ contains an orthonormal basis $\{\xi^j\}$, we can construct the matrix X and prove Lemma 2.1.2, which we used for proving Theorem 2.2.1.

However, we can easily verify that Theorem 2.2.1 holds also in the case that $N_{\tilde{a}} = \{0\}$, when everything becomes trivial. Indeed, we have

$$N_{\tilde{a}} = \{0\} \Leftrightarrow N_{\hat{a}} = \{0\}. \qquad (*)$$

Now, assume that, for some network $\hat{n}$, $N_{\tilde{a}} = \{0\}$. Then, if $i \in H^{c_2}$ is a solution of $\hat{n}$ corresponding to some $e \in H^{c_2}$, it follows from (*) and K_2 that necessarily $i = 0$ and i is unique. (Note that K_1 is always satisfied, since $N_{\hat{a}}^{\perp} = H^{c_2}$.) On the other hand, K_2^+ shows that the classical solution i' must also be zero; K_1^+ is satisfied, since the only $\gamma \in R^{c_2}$ satisfying $a\cdot\gamma = 0$ is zero. Hence, $i = i' = 0$.

To confirm the equivalence (*), we can proceed as follows: (a) Let $N_{\tilde{a}} \neq \{0\}$; then there exists $\eta \in R^{c_2}$, $\eta = [\eta_k] \neq 0$ such that $a\cdot\eta = 0$. Choose nonzero $h \in H$ and let $x = [\eta_k h]$; then clearly $x \in H^{c_2}$, $x \neq 0$ and $\hat{a}x = a\cdot x = 0$, i.e., $N_{\hat{a}} \neq \{0\}$.

(b) Conversely, assume that $N_{\hat{a}} \neq \{0\}$; then there exists $x = [x_i] \neq 0$, $x \in H^{c_2}$ such that $a\cdot x = 0$. Let $\{f_j\}$ be an orthonormal basis in H. Then, as in the proof of (ii) in Lemma 2.1.2, we see that there exist c_2-vectors $s^k \in R^{c_2}$ such that $x = \sum_k(s^k f_k)$, the series being convergent in H^{c_2}, and at least one of the s^k's, say s^{k*}, is nonzero. From this we conclude as before that $a\cdot s^k = 0$ for any k, and consequently, $a\cdot s^{k*} = 0$; hence $N_{\tilde{a}} \neq \{0\}$, which proves our equivalence (*).

A comment on the significance of Theorem 2.2.1 is in order. The relations K_1^+ and K_2^+ defining the classical solution are clearly straightforward extensions of classical Kirchhoff laws, i.e., in the case of a finite network, K_1^+ and K_2^+ coincide with the loop- and node-law, respectively. Thus, the classical solution appears to be the most natural concept of a solution. However, since the classical solution of a Hilbert network $\hat{n}$ coincides with the solution of the associated abstract network, all results established in Part 1 apply immediately to Hilbert networks.

Also, let us point out the following important fact: the Theorem 2.2.1 is in fact a generalization of Tellegen's Theorem [6] known from the classical network theory. Indeed, accepting the fact that the solution of a Hilbert network is defined by Definition 2, then Theorem 2.2.1 claims that the solution satisfies condition K_1, i.e., $\langle c, v-e\rangle = 0$ for every $c \in N_{\hat{a}}$; this, however, is Tellegen's theorem provided $\hat{n}$ is a finite network.

In the theorems on Hilbert networks that follow we will deal with mappings defined on a subset of H^{c_o} rather than with mappings on $N_{\hat{a}} \cap D$. To this end, let us introduce the following concepts:

If $D \subset H^{c_2}$ and $N_{\hat{a}} \cap D \neq \emptyset$, let $F \subset H^{c_o}$ be defined by

$$\hat{X}F = N_{\hat{a}} \cap D; \tag{2.2.6}$$

clearly, F is defined uniquely, since $\hat{X}H^{c_o} = N_{\hat{a}}$ and $\hat{X}$ is 1-to-1. If $N_{\hat{a}} \cap D = \emptyset$, we will put $F = \emptyset$. Due to Lemma 2.1.2 (v) we see that $F = \hat{X}^*(N_{\hat{a}} \cap D)$. Also, $F \subset (\hat{X}^* N_{\hat{a}}) \cap (\hat{X}^* D) = H^{c_o} \cap \hat{X}^* D = \hat{X}^* D$.

Moreover, recalling (1.2.2) and (ii) in Lemma 2.1.2, we let

$$Q(D) = (\hat{X}H^{c_o})^{\perp} + (\hat{Z}\hat{X}F)^{o} \tag{2.2.7}$$

for $F \neq \emptyset$, and $Q(D) = \emptyset$ for $F = \emptyset$.

<u>Theorem 2.2.2</u>. Let $\hat{n} = (\hat{Z}, G)$ be a H.n., and let $e \in H^{c_2}$; then $\hat{n}$ possesses a solution i corresponding to e, if and only if $e \in Q(D)$. In this case, the set I of all solutions corresponding to e is given by $I = Ae$, where $A : Q(D) \to \mathfrak{S}(\hat{X}F)$ is the admittance of $\hat{n}$. Moreover,

$$A = \hat{X}W^{-}\hat{}X^*, \tag{2.2.8}$$

where W^- signifies the quasi-inverse of the set mapping $W = \hat{X}^*\hat{Z}\hat{X} : F \to \mathfrak{S}((WF)^{o})$.

<u>Proof</u>: The first claim follows immediately from Theorem 1.2.1 and (2.2.7). To prove the second one, choose arbitrarily $e \in Q(D)$ and let

$$I' = Ae = \hat{X}W^{-}\hat{X}^{*}e. \tag{2.2.9}$$

First, we are going to show that the set I' is well-defined. Indeed, if $e \in Q(D)$, then, by (2.2.7), there exists $m \in (\hat{X}H^{c_o})^{\perp} = N_{\hat{a}}^{\perp} = N_{\hat{X}^*}$ and $n \in (\hat{Z}\hat{X}F)^{o}$ such that $e = m+n$. Since $\hat{X}^{*}m = 0$, we have

$$\hat{X}^{*}e = \hat{X}^{*}n \in \hat{X}^{*}(\hat{Z}\hat{X}F)^{o} = (WF)^{o}; \tag{2.2.10}$$

hence, $\hat{X}^{*}e$ is in the domain of W^{-}, and consequently, the expression $\hat{X}W^{-}\hat{X}^{*}e$ is meaningful.

Next, let $i \in I$. Then, by K_2, $i \in N_{\hat{a}} \cap D = \hat{X}F$, i.e., there exists $u \in F$ such that $i = \hat{X}u$. Moreover, by K_1, there exists $v \in \hat{Z}i$ such that $v-e \in N_{\hat{a}}^{\perp} = N_{\hat{X}^*}$; thus,

$$\hat{X}^{*}e = \hat{X}^{*}v. \tag{2.2.11}$$

Also, $\hat{X}^{*}v \in \hat{X}^{*}\hat{Z}i = \hat{X}^{*}\hat{Z}\hat{X}u = Wu$; this yields $u \in W^{-}\hat{X}^{*}v = W^{-}\hat{X}^{*}e$. Consequently, $i = \hat{X}u \in \hat{X}W^{-}\hat{X}^{*}e = I'$. Hence, $I \subset I'$.

Conversely, let $i \in I'$. Then, by (2.2.9), there exists $p \in W^{-}\hat{X}^{*}e$ such that $i = \hat{X}p$. Since $W^{-}\hat{X}^{*}e \subset F$, it follows that $p \in F$, and consequently, $i \in \hat{X}F = N_{\hat{a}} \cap D$, i.e., the element i satisfies K_2. On the other hand, the above relation yields $\hat{X}^{*}e \in Wp = \hat{X}^{*}\hat{Z}\hat{X}p = \hat{X}^{*}\hat{Z}i$; hence, there exists $v \in \hat{Z}i$ such that (2.2.11) holds. Thus, $v-e \in N_{\hat{X}^*} = N_{\hat{a}}^{\perp}$, i.e., i satisfies K_1. Hence, $i \in I$ so that $I' \subset I$ which concludes the proof.

<u>Theorem 2.2.3</u>. Let $\hat{n} = (\hat{Z},G)$ be a H.n., and let $F \neq \emptyset$; then $\hat{n}$ is regular on D, if and only if the set mapping $W = \hat{X}^{*}\hat{Z}\hat{X}$ is simple on F.

<u>Proof</u>: Denote $Y : N_{\hat{a}} \cap D \to F$ the inverse of $\hat{X} : F \to N_{\hat{a}} \cap D$; also note that $P = \hat{X}\hat{X}^{*}$ by Lemma 2.1.2 (vii). Then we have by Lemma 1.1.3: W is simple on $F \Leftrightarrow \hat{X}^{*}\hat{Z} = WY$ is simple on $N_{\hat{a}} \cap D \Leftrightarrow P\hat{Z} = \hat{X}\hat{X}^{*}\hat{Z}$ is simple on $N_{\hat{a}} \cap D$. This and Theorem 1.2.2 prove our statement.

<u>Theorem 2.2.4</u>. Let $\hat{n} = (\hat{Z},G)$ be a H.n., let $F \neq \emptyset$ and $W = \hat{X}^{*}\hat{Z}\hat{X}$; then $Q(D) = H^{c_2}$, if and only if $(WF)^{o} = H^{c_o}$.

<u>Corollary</u>: If $D \supset N_{\hat{a}}$, then $Q(D) = H^{c_2}$, if and only if $(WH^{c_o})^{o} = H^{c_o}$.

<u>Proof</u>: Using Theorem 1.2.3, (2.2.6) and Lemma 2.1.2 (ii), (vii), we have $Q(D) = H^{c_2} \Leftrightarrow [(P\hat{Z})(N_{\hat{a}} \cap D)]^{o} = N_{\hat{a}} \Leftrightarrow [\hat{X}\hat{X}^{*}\hat{Z}\hat{X}F]^{o} = \hat{X}H^{c_o} \Leftrightarrow \hat{X}(WF)^{o} = \hat{X}H^{c_o} \Leftrightarrow (WF)^{o} = H^{c_o}$.

For proving the corollary, it suffices to note that, by (2.2.6), $F = H^{c_o}$

At this point let us explain the reasons why in the above theorems we work with the set mapping W on F rather than with the set mapping $P\hat{Z}$ on $N_{\hat{a}} \cap D$. This approach is motivated by the most important case when the network is finite. In this case, we usually have $D = D_o^1 \times D_o^2 \times \ldots \times D_o^{c_2}$, where $D_o^i \subset H$ for $i = 1,2,\ldots,c_2$. Then it is easy to verify that $F = \tilde{D}_o^1 \times \tilde{D}_o^2 \times \ldots \times \tilde{D}_o^{c_o}$, where each set $\tilde{D}_o^k$ is a certain "linear combination" of the sets $D_o^1,\ldots,D_o^{c_2}$. (To see this, it suffices to realize that (2.2.6) yields $F = \hat{X}^*(N_{\hat{a}} \cap D)$.) However, it is easier to establish properties of a mpping on a set having the form $\tilde{D}_o^1 \times \ldots \times \tilde{D}_o^{c_o}$ than on a set $N_{\hat{a}} \cap D$. The advantage is even more striking in the case that $D_o^i = H$ for $i = 1,2,\ldots,c_2$; then clearly $F = H^{c_o}$.

<u>Theorem 2.2.5</u>. Let $\hat{n} = (\hat{Z},G)$ be a H.n., let $F \neq \emptyset$ and let $W = \hat{X}^*\hat{Z}\hat{X} : F \to \mathfrak{S}((WF)^o)$.

(a) If for all $y_1,y_2 \in F$, $y_1 \neq y_2$ and all $w_1 \in Wy_1$, $w_2 \in Wy_2$ we have

$$\langle w_1-w_2,y_1-y_2\rangle_{c_o} \neq 0, \tag{2.2.12}$$

then $\hat{n}$ is regular on D. If, in addition, $\hat{Z}$ is an operator, then

$$\langle Ae_1-Ae_2,e_1-e_2\rangle_{c_2} \neq 0 \tag{2.2.13}$$

for all $e_1,e_2 \in Q(D)$ such that $\hat{X}^*e_1 \neq \hat{X}^*e_2$, where A is the admittance of $\hat{n}$.

(b) If there exist constants $c > 0$ and $p > 1$ such that for any $y_1,y_2 \in F$ and any $w_1 \in Wy_1$, $w_2 \in Wy_2$ we have

$$|\langle w_1-w_2,y_1-y_2\rangle_{c_o}| \geq c\|y_1-y_2\|_{c_o}^p, \tag{2.2.14}$$

then $\hat{n}$ is regular on D; moreover,

$$\|Ae_1-Ae_2\|_{c_2} \leq c^{-\frac{1}{p-1}}\|\hat{X}^*(e_1-e_2)\|_{c_o}^{\frac{1}{p-1}} \tag{2.2.15}$$

and

$$|\langle Ae_1-Ae_2,e_1-e_2\rangle_{c_2}| \quad c^{-\frac{1}{p-1}}\|\hat{X}^*(e_1-e_2)\|_{c_o}^{\frac{p}{p-1}} \tag{2.2.16}$$

for all $e_1,e_2 \in Q(D)$.

(c) If there exist constants $c > 0$ and $p > 1$ such that for any $y_1,y_2 \in F$ and any $w_1 \in Wy_1$, $w_2 \in Wy_2$ we have

$$\mathrm{Re}\,\langle w_1-w_2,y_1-y_2\rangle_{c_o} \geq c\|y_1-y_2\|_{c_o}^p, \tag{2.2.17}$$

then $\hat{n}$ is regular on D, (2.2.15), (2.2.16) hold and

$$\mathrm{Re}\langle Ae_1-Ae_2,e_1-e_2\rangle_{c_2} \geq 0 \tag{2.2.18}$$

for all $e_1,e_2 \in Q(D)$.

Proof: (a) Choose $x_1,x_2 \in N_{\hat{a}}\cap D$ and $z_1 \in \hat{Z}x_1$, $z_2 \in \hat{Z}x_2$; since $\hat{X}$ is a 1-to-1 correspondence between F and $N_{\hat{a}}\cap D$, there exist unique elements $y_1,y_2 \in F$ such that $x_1 = \hat{X}y_1$, $x_2 = \hat{X}y_2$. Thus, we have

$$\begin{aligned}\langle z_1-z_2,x_1-x_2\rangle_{c_2} &= \langle z_1-z_2,\hat{X}(y_1-y_2)\rangle_{c_2} \\ &= \langle \hat{X}^*z_1-\hat{X}^*z_2,y_1-y_2\rangle_{c_0},\end{aligned} \tag{2.2.19}$$

and $\hat{X}^*z_j \in \hat{X}^*\hat{Z}x_j = \hat{X}^*\hat{Z}\hat{X}y_j = Wy_j$, $j = 1,2$. Hence, if $x_1 \neq x_2$, then $y_1 \neq y_2$, and consequently, by (2.2.19), $\langle z_1-z_2,x_1-x_2\rangle_{c_2} \neq 0$. Thus, by Theorem 1.2.4 (a), $\mathfrak{n}$, and $\hat{\mathfrak{n}}$ too, is regular on D. As for (2.2.13), it suffices to note that $Pe_1 = \hat{X}\hat{X}^*e_1 \neq \hat{X}\hat{X}^*e_2 = Pe_2 \Leftrightarrow \hat{X}^*e_1 \neq \hat{X}^*e_2$; then Theorem 1.2.4 (a) completes the proof.

The proof of (b) and (c) follows immediately from (b), (c) in Theorem 1.2.4 by using the equality (2.2.19) and the fact that $\hat{X}$ is an isometry between F and $N_{\hat{a}}\cap D$ (see Lemma 2.1.2 (iv)), i.e., $\|x_1-x_2\|_{c_2} = \|y_1-y_2\|_{c_0}$ whenever $x_j = \hat{X}y_j$, $j = 1,2$.

Further theorems are obtained by translating theorems in Part 1 into the language of Hilbert networks. We thus have the following:

Theorem 2.2.6. Let H be a real Hilbert space and let $\hat{\mathfrak{n}} = (\hat{Z},G)$ be a Hilbert network. Assume that

(i) $D \supset N_{\hat{a}}$

(ii) the set mapping $W = \hat{X}^*\hat{Z}\hat{X}$ is maximal montone and coercive on H^{c_0}.

Then $Q(D) = H^{c_2}$, i.e., for any $e \in H^{c_2}$, the network $\hat{\mathfrak{n}}$ possesses a solution corresponding to e.

The proof follows readily from Theorem 1.2.6 by using the following facts:

(a) $P\hat{Z}$ is maximal monotone on $N_{\hat{a}} \Leftrightarrow W$ is maximal monotone on H^{c_0}.

(b) $P\hat{Z}$ is coercive on $N_{\hat{a}} \Leftrightarrow W$ is coercive on H^{c_0}.

Note that monotonicity in (a) is guaranteed by (2.2.19), and maximality by the fact that $\hat{X}$ is 1-to-1. Similarly, (b) follows from the equality $\langle z,x\rangle_{c_2} = \langle w,y\rangle_{c_0}$ whenever $x = \hat{X}y$, $z \in \hat{Z}x$, $y \in H^{c_0}$ and $w = \hat{X}^*z \in \hat{X}^*\hat{Z}\hat{X}y = Wy$.

Theorem 2.2.7. Let H be a real Hilbert space and let $\hat{\eta} = (\hat{Z},G)$ be a H.n. Assume that

(i*) $D \supset N_{\hat{a}}$,

(ii*) there exist constants $c > 0$ and $p > 1$ such that for any $y_1,y_2 \in F$ and any $w_1 \in Wy_1$, $w_2 \in Wy_2$ with $W = \hat{X}^*\hat{Z}\hat{X}$ we have

$$\langle w_1-w_2,y_1-y_2\rangle_{c_o} \geq c\|y_1-y_2\|^p_{c_o} , \qquad (2.2.20)$$

(iii*) W is maximal monotone on H^{c_o}.

Then, for any $e \in H^{c_2}$, $\hat{\eta}$ possesses a unique solution corresponding to e.

The proof follows trivially from Theorem 1.2.7 and the above equivalence (a).

Problems

2.2.1 Let $\hat{\eta} = (\hat{Z},G)$ be a finite H.n., let $\hat{Y}$ have the same meaning as in Problem 2.1.3, and let the set $F' \subset H^{c_o}$ be defined by $\hat{Y}F' = N_{\hat{a}} \cap D$. S ow that Theorems 2.2.2-2.2.7 remain true, if everywhere $\hat{X}$ is replaced by $\hat{Y}$, F by F' and W by $W' = \hat{Y}^*\hat{Z}\hat{Y}$.

2.3 Quasi-linear and linear Hilbert networks

As in Section 1.3, here we will assume that $\hat{Z}$ of a H.n. $\hat{\eta}$ is an operator defined on the entire space H^{c_2}.

Theorem 2.3.1. Let $\hat{\eta} = (\hat{Z},G)$ be a H.n. with $\hat{Z}$ being an operator defined on H^{c_2}, and let $W = \hat{X}^*\hat{Z}\hat{X} : H^{c_o} \to H^{c_o}$. Assume that there exist constants $c > 0$ and $\lambda > 0$ such that

(i) $\mathrm{Re}\langle Wx_1-Wx_2,x_1-x_2\rangle_{c_o} \geq c\|x_1-x_2\|^2_{c_o}$ for all $x_1,x_2 \in H^{c_o}$,

(ii) $W \in \mathrm{Lip}_\lambda[H^{c_1},H^{c_o}]$.

Then $Q(H^{c_2}) = H^{c_2}$ and $\hat{\eta}$ is regular on H^{c_2}; moreover, the admittance A of $\hat{\eta}$ is in $\mathrm{Lip}_\gamma[H^{c_2},H^{c_2}]$ with $\gamma = (c\lambda)^{-1}(\lambda+\sqrt{\lambda^2-c^2})$.

Proof: Referring to Theorem 1.3.1, choose $x_j \in N_{\hat{a}}$, j = 1,2; then $x_j = \hat{X}y_j$ for some $y_j \in H^{c_o}$ and we have by (i),

$$\begin{aligned}\mathrm{Re}\langle\hat{Z}x_1-\hat{Z}x_2,x_1-x_2\rangle_{c_2} &= \mathrm{Re}\langle\hat{Z}\hat{X}y_1-\hat{Z}\hat{X}y_2,\hat{X}(y_1-y_2)\rangle_{c_2} \qquad (2.3.1)\\ &= \mathrm{Re}\langle Wy_1-Wy_2,y_1-y_2\rangle_{c_o} \geq c\|y_1-y_2\|^2_{c_o} .\end{aligned}$$

Since $\|y_1-y_2\|_{c_o} = \|x_1-x_2\|_{c_2}$ by Lemma 2.1.2 (iv), condition (i) in Theorem

1.3.1 is satisfied.

Moreover, for the above x_j, y_j we have by (ii),

$$\|P\hat{Z}x_1 - P\hat{Z}x_2\|_{c_2} = \|\hat{X}\hat{X}^*(\hat{Z}\hat{X}y_1 - \hat{Z}\hat{X}y_2)\|_{c_2} \tag{2.3.2}$$
$$= \|Wy_1 - Wy_2\|_{c_o} \leq \lambda\|y_1 - y_2\|_{c_o} = \lambda\|x_1 - x_2\|_{c_2},$$

i.e., (ii) in Theorem 1.3.1 is met; hence, the proof.

Referring to the Remark following Theorem 1.3.1 we see that conditions (i), (ii) in Theorem 2.3.1 can be replaced by stronger requirements, i.e.,

(i)' $\langle\hat{Z}x_1 - \hat{Z}x_2, x_1 - x_2\rangle_{c_o} \geq c\|x_1 - x_2\|^p_{c_2}$

for all $x_1, x_2 \in H^{c_2}$,

(ii)' $\hat{Z} \in Lip_\lambda[H^{c_2}, H^{c_2}]$.

Theorem 2.3.2. Let H be a real Hilbert space, let $\hat{n} = (\hat{Z}, G)$ be a H.n. with $\hat{Z}$ being an operator defined on H^{c_2}, and let $W = \hat{X}^*\hat{Z}\hat{X} : H^{c_o} \to H^{c_o}$. Assume that

(i)' there exist constants $c > 0$ and $p > 1$ such that

$$\langle Wx_1 - Wx_2, x_1 - x_2\rangle_{c_o} \geq c\|x_1 - x_2\|^p_{c_o} \tag{2.3.2}$$

for all $x_1, x_2 \in H^{c_o}$,

(ii)' W is continuous on H^{c_o}.

Then $Q(H^{c_2}) = H^{c_2}$ and $\hat{n}$ is regular on H^{c_2}; moreover, the admittance A of $\hat{n}$ satisfies the inequality

$$\|Ae_1 - Ae_2\|_{c_2} \leq c^{-\frac{1}{p-1}}\|\hat{X}^*(e_1 - e_2)\|^{\frac{1}{p-1}}_{c_o} \tag{2.3.4}$$

for all $e_1, e_2 \in H^{c_2}$.

Proof: Using the same reasoning as in the proof of Theorem 2.3.1, we confirm readily that our condition (i)' implies (1.3.3) in Theorem 1.3.2. Moreover, since $N_{\hat{a}} = \hat{X}H^{c_o}$ and $\hat{X}$ is 1-to-1 and continuous, the inverse $Y : N_{\hat{a}} \to H^{c_o}$ of $\hat{X}$ is also continuous according to the open mapping theorem. Thus, we have due to (ii)', $W = \hat{X}^*\hat{Z}\hat{X}$ is continuous on $H^{c_o} \Rightarrow \hat{X}^*\hat{Z} = WY$ is continuous on $N_{\hat{a}} \Rightarrow \hat{X}\hat{X}^*\hat{Z} = P\hat{Z}$ is continuous on $N_{\hat{a}}$; hence condition (ii)' in Theorem 1.3.2 is met, and our assertion is proved.

For proving (2.3.4) it suffices to realize that, for any $e_1, e_2 \in H^{c_2}$,

$\|P(e_1-e_2)\|_{c_2} = \|\hat{X}\hat{X}*(e_1-e_2)\|_{c_2} = \|\hat{X}*(e_1-e_2)\|_{c_o}$, and use inequality (1.2.8).

Remark 1. Let us stress the following important fact: Suppose that the operator $\hat{Z} : H^{c_2} \to H^{c_2}$ is described by a $c_2 \times c_2$ matrix $[Z_{ik}]$, where $Z_{ik} : H \to H$, i.e., $\hat{Z}$ is defined by (2.2.1). If the Z_{ik}'s are nonlinear, then the operator $\hat{X}*\hat{Z}\hat{X}$, appearing in Theorem 2.3.1 and 2.3.2, need not be equal to the operator described by the matrix $\bar{X}^T\cdot([Z_{ik}]\cdot X)$ (here, by $[Z_{ik}]\cdot X$ we understand the matrix obtained by formal multiplication of $[Z_{ik}]$ by X), because clearly $([Z_{ik}]\cdot X)\cdot x \neq [Z_{ik}]\cdot(X\cdot x)$, even if the network is finite. However, if all the operators Z_{ik}'s are linear, then such a representation is possible.

Let us now consider the case of a quasilinear network.

Theorem 2.3.3. Let $\hat{n} = (\hat{Z},G)$ be a H.n. and let $\hat{Z}$ be an operator defined on H^{c_2}. Assume that $\hat{Z} = \hat{Z}_1+\hat{Z}_2$, where $\hat{Z}_1 \in [H^{c_2},H^{c_2}]$ and let $W_j = \hat{X}*\hat{Z}_j\hat{X}$, $j = 1,2$. If

(i) $\inf_{\substack{\|x\|=1 \\ H^{c_o}}} \sup_{\substack{\|y\|=1 \\ H^{c_o}}} |\langle W_1x,y\rangle_{c_o}| = c > 0,$

$\inf_{\substack{\|y\|=1 \\ H^{c_o}}} \sup_{\substack{\|x\|=1 \\ H^{c_o}}} |\langle W_1x,y\rangle_{c_o}| > 0,$

(ii) there exists $\lambda > 0$ such that $W_2 \in Lip_\lambda[H^{c_o},H^{c_o}]$,

(iii) $c^{-1}\lambda < 1$,

then $Q(H^{c_2}) = H^{c_2}$, $\hat{n}$ is regular on H^{c_2} and the admittance A of $\hat{n}$ is in $Lip_\delta[H^{c_2},H^{c_2}]$, where $\delta = (c-\lambda)^{-1}$.

Proof: Choose $x,y \in N_{\hat{a}}$; then $x = \hat{X}x'$ and $y = \hat{X}y'$ for some $x',y' \in H^{c_o}$ and we have $\langle\hat{Z}_1x,y\rangle_{c_2} = \langle\hat{Z}_1\hat{X}x',\hat{X}y'\rangle_{c_2} = \langle W_1x',y'\rangle_{c_o}$. Since the spaces $N_{\hat{a}}$ and H^{c_o} are in a 1-to-1 norm preserving correspondence by Lemma 2.1.2 (ii) and (iv), conditions (i) imply conditions (i) in Theorem 1.3. Moreover, due to (2.3.2), our condition (ii) implies (ii) in Theorem 1.3.3, and Theorem 1.3.3 concludes the proof.

Theorem 2.3.4. Let $\hat{n} = (\hat{Z},G)$ be a H.n. , let $W = \hat{X}*\hat{Z}\hat{X}$, and assume that $W \in [H^{c_o},H^{c_o}]$, (or $Z \in [H^{c_2},H^{c_2}]$). Then

(a) $Q(H^{c_2}) = H^{c_2}$ and $\hat{n}$ is regular on H^{c_2}, if and only if W satisfies the conditions (i) in Theorem 2.3.3.

(b) If there exists a $c > 0$ such that

$$|\langle Wx,x\rangle_{c_o}| \geq c\|x\|^2_{c_o} \tag{2.3.5}$$

for all $x \in H^{c_o}$,

then $Q(H^{c_2}) = H^{c_2}$ and $\hat{n}$ is regular on H^{c_2}. In either case, the admittance A of $\hat{n}$ is in $[H^{c_2},H^{c_2}]$ and $\|A\| \leq c^{-1}$.

Proof: Using the same argument as in the proof of Theorem 2.3.2, we confirm that $W \in [H^{c_o},H^{c_o}] \Rightarrow P\hat{Z} \in [N_{\hat{a}},N_{\hat{a}}]$. The rest follows from Theorem 1.3.4.

Note that due to the Remark 3 following Theorem 1.3.4, the condition (2.3.5) can be replaced by the stronger requirement $|\langle\hat{Z}x,x\rangle_{c_o}| \geq c\|x\|^2_{c_2}$ for all $x \in H^{c_2}$; in this case, $\hat{n}$ is clearly regular independently of the structure of G.

2.4 Equivalent Definition of a Solution of a Hilbert Network

Using the ideas developed in Section 1.4, let us now establish an equivalent formulation of the solution concept for a Hilbert network. However, as we will see below, such a formulation is possible only for finite Hilbert networks.

As a matter of fact, if we wish to apply the theorems in Section 1.4, we need the essential assumption that $a^*\mathcal{H}'$ is closed in $\mathcal{H}$, i.e, in the case of a Hilbert network, that $\hat{a}^*H^{c_1}$ is closed in H^{c_2}. However, this assumption need not be satisfied, if the Hilbert network is infinite, i.e., if $c_1 = c_2 = \aleph_o$ (see Problem 2.4.1). On the other hand, everything is in order, if $c_1,c_2 < \aleph_o$. To see this, we will need the following proposition:

Lemma 2.4.1. Let $L \neq \{0\}$ be a normed linear space, let c,c' be positive integers, let A be a $c' \times c$ matrix with scalar elements, and let the operator $\hat{A} : L^c \to L^{c'}$ be defined by

$$\hat{A}x = A\cdot x; \tag{2.4.1}$$

then $\hat{A}L^c$ is closed in $L^{c'}$.

Remark 1. We understand that the topology in L^c (and in $L^{c'}$) is generated by the norm $\|x\|_c = (\sum_{k=1}^{c}\|x_k\|^2)^{\frac{1}{2}}$, $x = [x_k] \in L^c$.

Proof: If $A = 0$, then $\hat{A}L^c = \{0\}$ and our claim is trivially satisfied. Thus, let $A \neq 0$.

Let $a^1, a^2, \ldots, a^c$ be the columns of A, and let $a^{i_1}, a^{i_2}, \ldots, a^{i_k}$, $1 \leq k \leq c$ be linearly independent columns of A such that each a^j is a linear combination of a^{i_m}'s. Put $A' = [a^{i_1}|a^{i_2}|\ldots|a^{i_k}]$; then rank $A' = k$, and $1 \leq k \leq c'$. Moreover, define $\hat{A}' : L^k \to L^{c'}$ by $\hat{A}'y = A'\cdot y$. Then it is obvious that

$$\hat{A}L^c = \hat{A}'L^k. \tag{2.4.2}$$

Let us now distinguish two cases: (1) $c'-k = 0$, (2) $c'-k > 0$.

If case (1) occurs, then A' is a nonsingular $k \times k$-matrix; then clearly $\hat{A}'L^k = L^k$, and since L^k is closed, our proposition holds.

(2) Assume now that $c'-k > 0$. Construct the $c' \times \ell$ matrix U, whose columns constitute a complete linearly independent set of solutions of $A'^T\cdot u = 0$. Thus, we have $A'^T\cdot U = 0$, and $\ell = c'$-rank $A'^T = c'-k$; also note that rank $U = \ell = c'-k$.

Denote $Y = U^T$; then $A'^T\cdot Y^T = 0$, i.e.,

$$Y\cdot A' = 0. \tag{2.4.3}$$

Moreover, we have the claim:

The columns of A' constitute a complete linearly independent set of solutions of $Y\cdot v = 0$ with $v \in R^{c'}$.

Indeed, columns of A' are linearly independent and satisfy the equation $Y\cdot v = 0$. Furthermore, equation $Y\cdot v = 0$ has c'-rank $Y = c'$-rank $U = c'-(c'-k) = k$ linearly independent solutions, which is exactly the number of columns in A'.

To conclude the proof, define the operator $\hat{Y} : L^{c'} \to L^{\ell}$ by

$$\hat{Y}u = Y\cdot u, \tag{2.4.4}$$

and let $N_{\hat{Y}} = \{x : x \in L^{c'}, \hat{Y}x = 0\}$. Then

$$N_{\hat{Y}} = \hat{A}'L^k. \tag{2.4.5}$$

Indeed, let $x \in \hat{A}'L^k$; then $x = A'\cdot y$ for some $y \in L^k$, so that $\hat{Y}x = Y\cdot(A'\cdot y) = (Y\cdot A')\cdot y = 0$ by (2.4.3). Hence, $x \in N_{\hat{Y}}$, i.e., $\hat{A}'L^k \subset N_{\hat{Y}}$.

Conversely, let $x \in N_{\hat{Y}}$, i.e., $Y\cdot x = 0$. Let $\mathfrak{B}$ be a basis in L. Then, for the chosen x, there exists finitely many $b_1, b_2, \ldots, b_r \in \mathfrak{B}$ and a scalar $c' \times r$ matrix Q such that $x = Q\cdot b$ with $b = [b_1, b_2, \ldots, b_r]^T$. Thus, we have $Y\cdot(Q\cdot b) = (Y\cdot Q)\cdot b = 0$, which yields $Y\cdot Q = 0$ due to the linear independence of b_i's. Hence, by the above, there exists a $k \times r$ matrix M such that $Q = A'\cdot M$. Hence, $x = (A'\cdot M)\cdot b = A'\cdot(M\cdot b)$, and since $M\cdot b \in L^k$, we have

$x \in \hat{A}'L^k$; thus, $N_{\hat{Y}} \subset \hat{A}'L^k$, and (2.4.5) is proven.

However, since $\hat{Y}$ is a bounded operator on $L^{c'}$, it follows that $N_{\hat{Y}}$ is closed in $L^{c'}$; hence, by (2.4.5) and (2.4.2), $\hat{A}L^c$ is closed in $L^{c'}$ what we wanted to show.

Consider now a finite H.n. $\hat{n} = (\hat{Z},G)$. Since d is a finite matrix, we can take the identity matrix for K when constructing the matrix a by (2.1.4); consequently, the operator $\hat{a} \in [H^{c_2},H^{c_1}]$ is defined by equation

$$\hat{a}x = d^T \cdot x. \tag{2.4.6}$$

From this it follows that the adjoint $\hat{a}^* \in [H^{c_1},H^{c_2}]$ satisfies the relation

$$\hat{a}^*y = d \cdot y. \tag{2.4.7}$$

Hence, by Lemma 2.4.1, $\hat{a}^*H^{c_1}$ is closed in H^{c_2} and we can employ theorems in Section 1.4. Before going into detail, let us explain the physical meaning of equation (1.4.2) in case of a finite H.n.

Suppose that, for some $e \in H^{c_2}$, our network $\hat{n}$ possesses a solution $i \in H^{c_2}$; then, by (2.4.7) and Theorem 1.4.1, there exists $u \in \hat{Z}i \subset H^{c_2}$ and $v \in H^{c_1}$ such that

$$u + d \cdot v = e. \tag{2.4.8}$$

Consider some branch $b_j \in \mathcal{B}$, and suppose that v_{m_j} and $v_{m'_j}$ is the terminal and initial node of b_j respectively; thus, by definition of d, $d_{jm_j} = 1$, $d_{jm'_j} = -1$ and $d_{jk} = 0$ for all $k \neq m_j, m'_j$. If $u = [u_k]$, $v = [w_\ell]$, then (2.4.8) yields

$$u_j + w_{m_j} - w_{m'_j} = e_j. \tag{2.4.9}$$

However, equation (2.4.9) shows that the element w_ℓ has the meaning of a potential assigned to the node v_e, i.e., the difference of potentials at the terminal and initial node of a branch b_j equals the difference between the voltage of a source in b_j and voltage drop in b_j. (See Figure 2.4.)

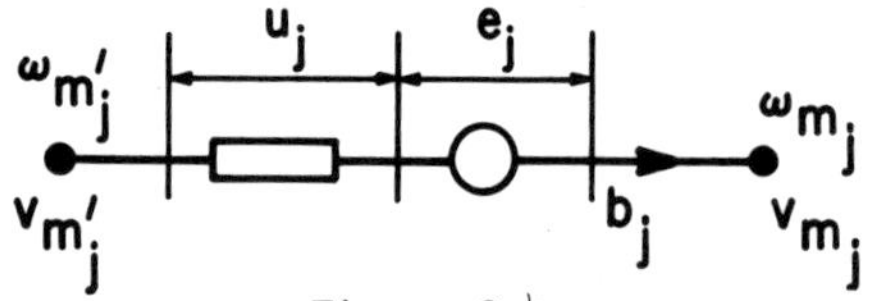

Figure 2.4

Since the translation of Theorems 1.4.1-1.4.3 into the language of Hilbert networks is straightforward, we omit the details. We present only a theorem resulting from Theorem 1.4.4.

Theorem 2.4.1. Let $\hat{n} = (\hat{Z},G)$ be a finite H.n. with $\hat{Z}$ being a linear operator defined on H^{c_2}, and let $d^T H^{c_2} = \{d^T\cdot x : x \in H^{c_2}\} \subset H^{c_1}$. Define the operator $M : H^{c_2} \times d^T H^{c_2} \to H^{c_2} \times d^T H^{c_2}$ by

$$M(x,y) = (\hat{Z}x + d\cdot y,\ d^T\cdot x). \tag{2.4.10}$$

Then $Q(H^{c_2}) = H^{c_2}$ and $\hat{n}$ is regular on H^{c_2} if and only if M possesses an inverse M^{-1}.

Proof: As mentioned above, we can set K = I. Then $\overline{\hat{a}H^{c_2}} = \overline{d^T H^{c_2}}$; moreover, by Lemma 1.4.1 and 1.4.2, $N^{\perp}_{\hat{a}^*} = (\hat{a}H^{c_2})^{\perp\perp} = \overline{(\hat{a}H^{c_2})} = \overline{d^T H^{c_2}}$. However, by Lemma 2.4.1, $d^T H^{c_2}$ is closed in H^{c_1}, and consequently, $N^{\perp}_{\hat{a}^*} = d^T H^{c_2}$. The rest follows from Theorem 1.4.4 and (2.4.6), (2.4.7).

Remark 2. Since the operator $\hat{Z}$ in Theorem 2.4.1 is assumed to be linear, it can be generated by a $c_2 \times c_2$ matrix Z whose entries are linear operators from H into itself. Thus, it follows from (2.4.10) that M can be generated by the matrix $\begin{bmatrix} Z & d \\ d^T & 0 \end{bmatrix}$.

Problems

2.4.1 Motivated by Lemma 2.4.1 one is tempted to believe that the following conjecture is true:

Let $A = [A_{ik}]$ be an infinite matrix such that $\sum_{i,k=1}^{\infty} |A_{ik}|^2 < \infty$, and let the operator $\hat{A} : R^{\aleph_0} \to R^{\aleph_0}$ $(R^{\aleph_0} = \ell_2)$ be defined by $\hat{A}x = A\cdot x$; then $\hat{A}R^{\aleph_0}$ is closed in $R^{\aleph_0}$.

Show that this conjecture is false.

2.5 Qualitative Properties of Hilbert Networks

In this section we are going to apply results obtained in Section 1.5 to Hilbert networks and thus derive theorems of a qualitative character.

Let the symbols $\mathfrak{P}_{\varphi}(\tilde{D})$, $\mathfrak{M}_{\varphi}(\tilde{D})$ have the same meaning as in Section 1.5; then we have

Theorem 2.5.1. Let $\hat{n} = (\hat{Z},G)$ be a Hilbert network with $F \neq \emptyset$, let $A : Q(D) \to N_{\hat{a}} \cap D$ be the admittance of $\hat{n}$, and let $W = \hat{X}^*\hat{Z}\hat{X}$; then

(i) $A \in \mathfrak{P}_\varphi(Q(D)) \Leftrightarrow W \in \mathfrak{P}_\varphi(F)$,

(ii) $A \in \mathfrak{M}_\varphi(Q(D)) \Leftrightarrow W \in \mathfrak{M}_\varphi(F)$,

where the set $F \subset H^{c_o}$ is defined by (2.2.6).

The proof follows trivially from Theorem 1.5.1 and the equality (2.2.19) which shows that $W \in \mathfrak{P}_\varphi(F) \Leftrightarrow \hat{Z} \in \mathfrak{P}_\varphi(N_a \cap D)$, and similarly for $\mathfrak{M}_\varphi$.

Note also that the Corollary following Theorem 1.5.1 holds without any change for Hilbert networks.

Theorem 2.5.2. Let $\hat{n} = (\hat{Z}, G)$ be a Hilbert network, and let $\hat{Z}$ be an operator on $D \subset H^{c_2}$. Assume that

(i) there exists $c > 0$ and $p > 1$ such that

$$\mathrm{Re}\langle Wx_1 - Wx_2, x_1 - x_2\rangle_{c_o} \geq c\|x_1 - x_2\|_{c_o}^{p} \tag{2.5.1}$$

for all $x_1, x_2 \in F$, where $W = \hat{X}^*\hat{Z}\hat{X}$,

(ii) there exists $d > 0$ and $q > 0$ such that

$$\|Wx_1 - Wx_2\|_{c_o} \leq d\|x_1 - x_2\|_{c_o}^{q} \tag{2.5.2}$$

for all $x_1, x_2 \in F$.

Then $\hat{n}$ is regular on D and the admittance A of $\hat{n}$ satisfies the inequality

$$\mathrm{Re}\langle Ae_1 - Ae_2, e_1 - e_2\rangle_{c_2} \geq c d^{-\frac{p}{q}} \|\hat{X}^*(e_1 - e_2)\|_{c_o}^{\frac{p}{q}} \tag{2.5.3}$$

for all $e_1, e_2 \in Q(D)$.

Proof: Referring to the equation (2.2.19) it follows readily that (2.5.1) implies the condition (i) in Theorem 1.5.2. Moreover, if $x_j \in N_a \cap D$, $j = 1,2$, then $x_j = \hat{X}y_j$ for some $y_j \in F$, and consequently, $\|P\hat{Z}x_1 - P\hat{Z}x_2\|_{c_2} = \|\hat{X}\hat{X}^*\hat{Z}\hat{X}y_1 - \hat{X}\hat{X}^*\hat{Z}\hat{X}y_2\|_{c_2} = \|Wy_1 - Wy_2\|_{c_o} \leq d\|y_1 - y_2\|_{c_o}^{q}$ by (2.5.2); since $\|x_1 - x_2\|_{c_2} = \|y_1 - y_2\|_{c_o}$, condition (1.5.5) follows. Finally, the fact that $\|Pe_1 - Pe_2\|_{c_2} = \|\hat{X}^*(e_1 - e_2)\|_{c_o}$ for $e_j \in Q(D)$ and Theorem 1.5.2 conclude the proof.

Let us now prove the least power theorem which generalizes the classical result of J. Jeans [7]. To this end, let us introduce the following concepts:

Let $s = (n_i)$ be a (finite or infinite) increasing sequence of positive integers. A c-vector $x = [x_k]$, $c \leq \aleph_o$, will be said to have the s-pattern, if $x_k = 0$ for all indices $k \notin s$; analogously, a c-vector $y = [y_k]$ will be said to have the s-copattern, if $y_k = 0$ for all $k \in s$.

Next, if N is a linear subspace of H^c, $c \leq \aleph_o$, then ${}^{c(s)}N$ will signify the linear subspace of H^c consisting of all vectors in N which have the s-copattern.

Then we have

Theorem 2.5.3. Let $\hat{n} = (\hat{Z},G)$ be a Hilbert network such that $\hat{Z}$ is a self-adjoint, strictly positive operator on H^{c_2} (i.e., $\langle\hat{Z}x,x\rangle_{c_2} > 0$ for every $x \in H^{c_2}$, $x \neq 0$). Furthermore, let $e \in H^{c_2}$ have an s-pattern, and let ${}^{c(s)}N_{\hat{a}} \neq \{0\}$. Then the solution $i \in H^{c_2}$ of $\hat{n}$ corresponding to e is determined uniquely and the functional

$$P(x) = \langle\hat{Z}x,x\rangle_{c_2} - \langle\hat{Z}i,i\rangle_{c_2} \qquad (2.5.4)$$

is positive on $i + {}^{c(s)}N_{\hat{a}}$ except for $x = i$.

Proof: Referring to definition (1.5.9) of the set F_e, we see readily that ${}^{c(s)}N_{\hat{a}} \subset F_e \cap N_{\hat{a}}$. Thus, by Theorem 1.5.3, the function P is positive on $i + {}^{c(s)}N_{\hat{a}} \subset i + F_e \cap N_{\hat{a}}$ except for the argument $x = i$. (Note that the self-adjointness and strict positivity of $\hat{Z}$ imply conditions (i) and (ii) in Theorem 1.5.3.) Hence, the proof.

An explanation concerning the physical meaning of Theorem 2.5.3 is in order. If $e \in H^{c_2}$ has an s-pattern, $s = (n_1,n_2,n_3,\ldots)$, then this fact means that voltage sources are present only in the branches $b_{n_i} \in \mathcal{B}$, $n_i \in s$. Now, if $i \in H^{c_2}$ is a solution of $\hat{n}$ corresponding to e, then the set $i + {}^{c(s)}N_{\hat{a}} \subset N_{\hat{a}}$ has the meaning of the set of all current regimes in $\hat{n}$ such that each regime $x \in i + {}^{c(s)}N_{\hat{a}}$ coincies with the regime i in branches b_{n_i}, $n_i \in s$, i.e., $x_{n_i} = i_{n_i}$ for each $n_i \in s$. Then, obviously, our result claims that the dissipated power $\langle\hat{Z}x,x\rangle_{c_2}$ in the network $\hat{n}$, when considering regimes $x \in i + {}^{c(s)}N_{\hat{a}}$, is minimized precisely when $x = i$. Hence, the solution i has the property that, from all current regimes described, i produces minimal dissipated power in $\hat{n}$.

Let us now prove an approximation theorem for a solution of an infinite network. To this end, we will need some graph-theoretic concepts.

Let G be an infinite $(c_2 = \aleph_o)$ locally finite graph, and let $\mathcal{B} = \{b_1,b_2,b_3,\ldots\}$ be the set of branches of G. For every $b_i \in \mathcal{B}$, let v_{i+} and v_{i-} be the terminal and initial node of b_i, respectively, i.e.,

$Gb_i = (v_{i-}, v_{i+})$.

If $n = 1$ is an integer, let G_n be the restriction of the mapping G to the set $\mathcal{B}_n = \{b_1, b_2, \ldots, b_n\}$. If we let

$$\mathcal{V}^{(n)} = \bigcup_{i=1}^{n} \{v_{i-}, v_{i+}\}, \qquad (2.5.5)$$

we see easily that G_n is actually a mapping from $\mathcal{B}_n$ into $\mathcal{V}^{(n)} \times \mathcal{V}^{(n)}$ which satisfies the requirements (i) and (ii) in the definition of an oriented graph; hence, in this interpretation, G_n is a (finite) oriented graph.

Moreover, if d is the incidence matrix of G, and d_n the incidence matrix of G_n, we see easily that d_n is obtained from d by deleting the $n+1$, $n+2$, $n+3, \ldots$ row, and by deleting all columns corresponding to nodes v_α with $v_\alpha \notin \mathcal{V}^{(n)}$, i.e. all nodes which do not enter into the graph G_n.

Next, if $x = [x_k]$ is an $\aleph_0$-vector, let $(x)_n$ signify the n-vector $[x_1, x_2, \ldots, x_n]^T$; similarly, if $y = [y_k]$ is an n-vector, let y' stand for the $\aleph_0$ vector $[y_1, y_2, \ldots, y_n, 0, 0, 0, \ldots]^T$.

Furthermore, let $\hat{n} = (\hat{Z}, G)$ be an infinite Hilbert network (i.e., $c_2 = \aleph_0$) and assume that $\hat{Z}$ is an operator from H^{c_2} into itself. If $n \geq 1$ is an integer, let $\hat{Z}_n$ be the operator from H^n into itself defined by

$$\hat{Z}_n x = (\hat{Z}x')_n, \qquad x \in H^n. \qquad (2.5.6)$$

Then clearly $\hat{n}_n = (\hat{Z}_n, G_n)$ is a (finite) subnetwork of $\hat{n}$. Moreover, if $e_n \in H^n$, then by a solution i_n of $\hat{n}_n$ we will understand, of course, an n-vector satisfying the relations K_1, K_2 applied for $\hat{n}_n$.

Note that, if in particular, the operator $\hat{Z}$ is described by a matrix $[Z_{ik}]$ and defined by (2.2.1), then $\hat{Z}_n$ is clearly described by the $n \times n$ submatrix of $[Z_{ik}]$ standing in the upper left corner.

Theorem 2.5.4. Let $\hat{n} = (\hat{Z}, G)$ be an infinite Hilbert network with $\hat{Z}$ being an operator on $H^{\aleph_0}$, let $W = \hat{X}^* \hat{Z} \hat{X} : H^{c_0} \to H^{c_0}$, and let $\hat{n}_n$ have the meaning defined above. Assume that the following conditions are satisfied:

(1) there exists constants $\gamma, \lambda > 0$ such that

$$\operatorname{Re} \langle Wx_1 - Wx_2, x_1 - x_2 \rangle_{c_0} \geq \gamma \|x_1 - x_2\|^2_{c_0} \qquad (2.5.7)$$

for all $x_1, x_2 \in H^{c_0}$, and

$$\hat{Z} \in \mathrm{Lip}_\lambda [H^{\aleph_0}, H^{\aleph_0}]. \qquad (2.5.8)$$

(2) The set S of all simple vectors in $N_{\hat{a}}$ is dense in $N_{\hat{a}}$.

Then,

(a) $\hat{\eta}$ is regular on $H^{\aleph_o}$, $Q(H^{\aleph_o}) = H^{\aleph_o}$, and each $\hat{\eta}_n$ is regular on H^n and $Q(H^n) = H^n$,

(b) if $i \in H^{\aleph_o}$ is the solution of $\hat{\eta}$ corresponding to some $e \in H^{\aleph_o}$ and $i_n \in H^n$ is the solution of $\hat{\eta}_n$ corresponding to $(e)_n \in H^n$, then $i'_n \to i$ in $H^{\aleph_o}$. If, in addition $\hat{Z}$ is linear, then conditions (2.5.7) and (2.5.8) may be replaced by the requirements

$$|\langle Wx,x\rangle_{c_o}| \geq \gamma\|x\|^2_{c_o} \qquad (2.5.9)$$

for all $x \in H^{c_o}$, and $\hat{Z} \in [H^{\aleph_o},H^{\aleph_o}]$.

Proof: For each integer $n \geq 1$, let $H^{n'}_{\aleph} = \{x' : x \in H^n\}$; then $H^{1'} \subset H^{2'} \subset H^{3'} \subset \ldots \subset H^{\aleph_o}$. Let $P_n : H^{\aleph_o} \to H^{n'}$ be defined by $P_n x = \{(x)_n\}'$; then clearly P_n is the orthogonal projection from $H^{\aleph_o}$ onto $H^{n'}$.

Next, let d_n be the incidence matrix of G_n. From the definition of the graph it follows that if $x \in H^n$ satisfies the equation $d^T_n \cdot x = 0$, then $d^T_{n+1} \cdot (x')_{n+1} = 0$ and $a \cdot x' = 0$. Thus, letting

$$N_{\hat{d}^T_n} = \{x : x \in H^n, d^T_n \cdot x = 0\}, \quad N'_{\hat{d}^T_n} = \{x' : x \in N_{\hat{d}^T_n}\},$$

we have

$$N'_{\hat{d}^T_n} \subset H^{n'}, \quad N'_{\hat{d}^T_n} \subset N'_{\hat{d}^T_{n+1}} \subset N_{\hat{a}}. \qquad (2.5.10)$$

Define now the $\aleph_o \times \aleph_o$ matrices a_n by

$$a_n = \left[\begin{array}{c|c} d^T_n & 0 \\ \hline 0 & \tilde{d} \end{array}\right], \qquad (2.5.11)$$

where $\tilde{d}$ is the $\aleph_o \times \aleph_o$ diagonal matrix

$$\tilde{d} = \mathrm{diag}(1,\tfrac{1}{2},\tfrac{1}{3},\tfrac{1}{4},\ldots), \qquad (2.5.12)$$

and let $\hat{a}_n \in [H^{\aleph_o},H^{\aleph_o}]$ be defined by $\hat{a}_n x = a_n \cdot x$. From (2.5.12) and (2.5.11) it follows readily that $N_{\hat{a}_n} = N'_{\hat{d}^T_n}$, where $N_{\hat{a}_n} \subset H^{\aleph_o}$ is the null space of $\hat{a}_n$. Hence, $N_{\hat{a}_n} \subset H^{n'}$ and $N_{\hat{a}_n} \subset N_{\hat{a}_{n+1}}$, i.e., the condition (ii) in Theorem 1.5.4 is satisfied.

Also, it is clear that assumption (2) is exactly the condition (iii) in Theorem 1.5.4.

On the other hand, as we have seen before, the condition (2.5.7) is equivalent to the condition

$$\mathrm{Re}\langle\hat{Z}y_1-\hat{Z}y_2,y_1-y_2\rangle_{\aleph_o} \geq \gamma\|y_1-y_2\|^2_{\aleph_o} \qquad (2.5.13)$$

for all $y_1, y_2 \in N_{\hat{a}}$. However, if P is the orthogonal projection of $H^{\aleph_0}$ onto $N_{\hat{a}}$, then, by (2.5.8),

$$P\hat{Z} \in \mathrm{Lip}_\lambda[N_{\hat{a}}, N_{\hat{a}}]. \tag{2.5.14}$$

Hence, by Theorem 1.3.1, the abstract network $(\hat{Z}, \hat{a})$ over $H^{\aleph_0}$ is regular on $H^{\aleph_0}$ and $Q(H^{\aleph_0}) = H^{\aleph_0}$, i.e., the same is true for $\hat{h}$.

Next, we are going to show that each network $\hat{h}_n$ is regular. Indeed, if $u_1, u_2 \in N_{\hat{d}^T_n} \subset H^n$, we have by (2.5.13), (2.5.6) and the inclusion $N_{\hat{d}^T_n} \subset N_{\hat{a}}$,

$$\begin{aligned} \mathrm{Re}\langle \hat{Z}_n u_1 - \hat{Z}_n u_2, u_1 - u_2\rangle_n &= \mathrm{Re}\langle (\hat{Z}u_1')_n - (\hat{Z}u_2')_n, u_1 - u_2\rangle_n \\ &= \mathrm{Re}\langle \hat{Z}u_1' - \hat{Z}u_2', u_1' - u_2'\rangle_{\aleph_0} \geq \gamma\|u_1' - u_2'\|^2_{\aleph_0} \\ &= \gamma\|u_1 - u_2\|_n^2. \end{aligned} \tag{2.5.15}$$

Furthermore, if Q_n is the orthogonal projection from H^n onto $N_{\hat{d}^T_n}$ and $u_1, u_2 \in N_{\hat{d}^T_n}$, then $u_1', u_2' \in N_{\hat{a}}$ and (2.5.9) yields $\|Q_n(\hat{Z}_n u_1 - \hat{Z}_n u_2)\|_n^n \leq \|\hat{Z}_n u_1 - \hat{Z}_n u_2\|_n = \|(\hat{Z}u_1')_n - (\hat{Z}u_2')_n\|_n \leq \|\hat{Z}u_1' - \hat{Z}u_2'\|_{\aleph_0} \leq \lambda\|u_1' - u_2'\|_{\aleph_0} = \lambda\|u_1 - u_2\|_n$, i.e.,

$$Q_n\hat{Z}_n \in \mathrm{Lip}_\lambda[N_{\hat{d}^T_n}, N_{\hat{d}^T_n}]. \tag{2.5.16}$$

Then, according to Theorem 1.3.1, (2.5.15) and (2.5.16) show that the abstract network $(\hat{Z}_n, \hat{d}^T_n)$ over H^n is regular on H^n, $Q(H^n) = H^n$, and consequently, the same is true for $\hat{h}_n$.

Consider now the abstract network $h_n = (\hat{Z}, \hat{a}_n)$ over $H^{\aleph_0}$. Since $N_{\hat{a}_n} = N'_{\hat{d}^T_n} \subset N_{\hat{a}}$, we get readily from (2.5.13)

$$\mathrm{Re}\langle \hat{Z}y_1 - \hat{Z}y_2, y_1 - y_2\rangle_{\aleph_0} \geq \gamma\|y_1 - y_2\|^2_{\aleph_0} \tag{2.5.17}$$

for all $y_1, y_2 \in N_{\hat{a}_n}$. Similarly, if K_n is the orthogonal projection from $H^{\aleph_0}$ onto $N_{\hat{a}_n}$, (2.5.8) implies as above that

$$K_n\hat{Z} \in \mathrm{Lip}_\lambda[N_{\hat{a}_n}, N_{\hat{a}_n}]. \tag{2.5.18}$$

Hence, (2.5.17), (2.5.18) show by Theorem 1.3.1 that h_n is regular and $Q(H^{\aleph_0}) = H^{\aleph_0}$.

Finally, we are going to show that if $i_n \in H^n$ is the solution of $\hat{h}_n$ corresponding to $(e)_n \in H^n$, then $i_n' \in H^{\aleph_0}$ is the solution of h_n corresponding to $\{(e)_n\}' = P_n e$. Indeed, we have $i_n \in N_{\hat{d}^T_n}$ and

$$\langle \tilde{c}, \hat{Z}_n i_n - (e)_n\rangle_n = 0 \tag{2.5.19}$$

for each $\tilde{c} \in N_{\hat{d}_T}^{n}$. Denote $J_n \in H^{\aleph_o}$ the solution of $\mathfrak{h}_n$ corresponding to $P_n e$; Then $J_n \in N_{\hat{a}_n}$ and

$$\langle c, \hat{Z}J_n - \{(e)_n\}' \rangle_{\aleph_o} = 0 \qquad (2.5.20)$$

for all $c \in N_{\hat{a}_n}$. However, due to the definition (2.5.11) of a_n it follows that there exists $I_n \in H^n$ such that $J_n = I_n'$, and by equality $a_n \cdot J_n = 0$ we necessarily have $I_n \in N_{\hat{d}_T}^{n}$. Thus, for any $c^+ \in N_{\hat{d}_T}^{n}$ we have $(c^+)' \in N_{\hat{a}_n}$, and (2.5.20) yields $0 = \langle (c^+)', \hat{Z}I_n' - \{(e)_n\}' \rangle_{\aleph_o} = \langle c^+, (\hat{Z}I_n')_n - (e)_n \rangle_n$ $= \langle c^+, \hat{Z}_n I_n - (e)_n \rangle_n$. Comparing this with (2.5.19) we see that, due to regularity of $\hat{\mathfrak{h}}_n$ and $\mathfrak{h}_n$, necessarily $i_n = I_n$; consequently, $J_n = i_n'$ what we wanted to show.

Applying now Theorem 1.5.4 to networks $\mathfrak{h}$ and $\mathfrak{h}_n$, we conclude that $i_n' \to i$ in $H^{\aleph_o}$ and the proof is complete.

From Theorem 1.5.5 we obtain immediately the following generalization:

Theorem 2.5.5. Let H be real; then Theorem 2.5.4 remains true, if the requirement (1) is replaced by the subsequent conditions (i) and (ii):

(i) there exist constants $\gamma > 0$ and $p \geq 2$ such that

$$\langle Wx_1 - Wx_2, x_1 - x_2 \rangle_{c_o} \geq \gamma \|x_1 - x_2\|_{c_o} \qquad (2.5.21)$$

for all $x_1, x_2 \in H^{c_o}$,

(ii) $\hat{Z}$ is continuous on $H^{\aleph_o}$.

The proof follows the same pattern as before and is omitted.

As a final result in this section, we establish a bound on the power dissipated in a linear Hilbert network when it is driven by an EMF vector $e \in H^{c_2}$. As we shall see, such a bound is $\langle \hat{Z}^{-1} e, e \rangle_{c_2}$ when $\hat{Z}$ is invertible. Furthermore, a result of this nature can be obtained under a weaker condition than the invertibility of $\hat{Z}$.

As before, let $s = (n_i)$ be a finite or infinite, increasing sequence of positive integers. ${}^{s}H^{c_2}$ and ${}^{c(s)}H^{c_2}$ will denote the subspaces of H^{c_2} consisting of all c_2-vectors $[x_k] \in H^{c_2}$ having the s-pattern and the s-co-pattern, respectively. Thus, $H^{c_2} = {}^{s}H^{c_2} \oplus {}^{c(s)}H^{c_2}$. Also, $\hat{Z}_s$ and $\hat{Z}_{c(s)}$ will denote the restrictions of $\hat{Z} \in [H^{c_2}, H^{c_2}]$ to ${}^{s}H^{c_2}$ and ${}^{c(s)}H^{c_2}$, respectively.

Theorem 2.5.6 (Zemanian [8]). Let $\hat{\mathfrak{h}} = (\hat{Z}, G)$ be a Hilbert network, and let

$\hat{Z} \in [H^{c_2}, H^{c_2}]$ be positive. Furthermore, let $e \in H^{c_2}$ have an s-pattern and assume that ${}^{s}H^{c_2}$ reduces $\hat{Z}$. Also, assume that $\hat{Z}_s$ is invertible on ${}^{s}H^{c_2}$. If $i \in H^{c_2}$ is a solution of $\hat{\eta}$ corresponding to e, then

$$0 \leq \langle i, \hat{Z} i \rangle_{c_2} \leq \langle \hat{Z}_s^{-1} e, e \rangle_{c_2}. \qquad (2.5.22)$$

Proof: Since $\hat{Z}$ is positive, so too are $\hat{Z}_s$ and $\hat{Z}_{c(s)}$. Moreover, $\hat{Z}_s$ possesses a positive square root $\hat{Z}_s^{\frac{1}{2}}$. Also, since $\hat{Z}_s^{\frac{1}{2}} \hat{Z}_s^{\frac{1}{2}} = \hat{Z}_s$ and $\hat{Z}_s$ is invertible on ${}^{s}H^{c_2}$, $\hat{Z}_s^{-\frac{1}{2}}$ exists as a positive operator from ${}^{s}H^{c_2}$ onto ${}^{s}H^{c_2}$. (By $\hat{Z}_s^{-\frac{1}{2}}$ we mean $(\hat{Z}_s^{\frac{1}{2}})^{-1}$).

Now, $i = i_s + i_{c(s)}$, where $i_s \in {}^{s}H^{c_2}$ and $i_{c(s)} \in {}^{c(s)}H^{c_2}$. Thus.

$$\langle i, e \rangle_{c_2} = \langle i_s, e \rangle_{c_2} = \langle \hat{Z}_s^{-\frac{1}{2}} \hat{Z}_s^{\frac{1}{2}} i_s, e \rangle_{c_2} = \langle \hat{Z}_s^{\frac{1}{2}} i_s, \hat{Z}_s^{-\frac{1}{2}} e \rangle_{c_2}.$$

By Schwartz inequality,

$$|\langle i, e \rangle_{c_2}|^2 \leq \|\hat{Z}_s^{\frac{1}{2}} i_s\|_{c_2}^2 \cdot \|\hat{Z}_s^{-\frac{1}{2}} e\|_{c_2}^2 \qquad (2.5.23)$$
$$= \langle i_s, \hat{Z}_s i_s \rangle_{c_2} \cdot \langle \hat{Z}_s^{-1} e, e \rangle_{c_2}.$$

Since ${}^{s}H^{c_2}$ reduces $\hat{Z}$, we have that $\langle i, \hat{Z} i \rangle_{c_2} = \langle i_s, \hat{Z}_s i_s \rangle_{c_2} + \langle i_{c(s)}, \hat{Z}_{c(s)} i_{c(s)} \rangle_{c_2}$. Since both terms on the right-hand side are nonnegative, we have

$$\langle i_s, \hat{Z}_s i_s \rangle_{c_2} \leq \langle i, \hat{Z} i \rangle_{c_2}.$$

Therefore, (2.5.23) yields

$$|\langle i, e \rangle_{c_2}|^2 \leq \langle i, \hat{Z} i \rangle_{c_2} \langle \hat{Z}_s^{-1} e, e \rangle_{c_2}. \qquad (2.5.24)$$

Finally, $\langle i, e \rangle_{c_2} = \langle i, \hat{Z} i \rangle_{c_2} \geq 0$ because i is a solution of $\hat{\eta}$ corresponding to e. Therefore, (1.5.25) yields (2.5.22).

It is worth pointing out that the right-hand side of (2.5.22), which is a bound on the power dissipation $\langle i, Zi \rangle_{c_2}$ in $\hat{\eta}$ is independent of the topology and the element values in that part of $\hat{\eta}$ corresponding to the s-copattern. Also, a special case of Theorem 2.5.5 arises when $\hat{Z}$ is described by a diagonal matrix $Z = \mathrm{diag}(Z_1, Z_2, \ldots)$, and each $Z_j \in [H,H]$ is invertible whenever $e_j \neq 0$. In this case, (2.5.22) reduces to

$$\langle i, \hat{Z} i \rangle_{c_2} \leq \sum_j \langle Z_j^{-1} e, e \rangle_1,$$

where the summation is over those indices j for which $e_j \neq 0$.

2.6 Causality of Hilbert Networks

In this section, we will give conditions for causality of the admittance of a Hilbert network.

The Theorems 1.6.1 and 1.6.3 can readily be applied to Hilbert networks; it suffices to realize that in this case $\mathcal{H} = H^{c_2}$ and, if necessary, relations (1.6.2) and (1.6.8) can be replaced by equivalent relations involving the set mapping $\hat{X}*\hat{Z}\hat{X} = W : F \to WF$. Since this replacing is trivial, we omit the details; instead, we are going to consider a particular type of causality of mappings on certain $D \subset H^{c_2}$, which is generated by a resolution of the identity on H.

Lemma 2.6.1. Let $\{E_T : T \in R^1\}$ be a resolution of identity on H, and let $c \leq \aleph_0$. For any $T \in R^1$, let $S_T : H^c \to H^c$ be defined by

$$S_T[x_k] = [E_T x_k], \; x = [x_k] \in H^c. \tag{2.6.1}$$

Then the collection $\{S_T : T \in R^1\}$ is a resolution of identity on H^c.

Proof: Clearly, S_T is a bounded linear operator from H^c into itself, since for any integer $N > 0$ and $x \in H^c$ we have $\sum_{k=1}^{N} \|E_T x_k\|^2 \leq \sum_{k=1}^{N} \|E_T\|^2 \|x_k\|^2 \leq \|x\|_c^2$. (Note that E_T, being an orthogonal projection, satisfies $\|E_T\| = 1$.)

Moreover, we can see easily that S_T is a projection. Indeed, for any $T \in R^1$ and $x \in H^c$, $S_T^2 x = [E_T^2 x] = [E_T x] = S_T x$, so that $S_T^2 = S_T$.

Also, if $x,y \in H^c$, we have $\langle S_T x, y\rangle_c = \sum_k \langle E_T x_k, y_k\rangle = \sum_k \langle x_k, E_T y_k\rangle = \langle x, S_T y\rangle_c$; thus, $S_T^* = S_T$.

Next, let $T_1 \leq T_2$; then, for any $x \in H^c$, $\langle (S_{T_2} - S_{T_1})x, x\rangle_c = \sum_k \langle E_{T_2} x_k - E_{T_1} x_k, x_k\rangle \geq 0$. Hence, $S_{T_2} \geq S_{T_1}$ and condition (i) in the definition of a resolution of identity is satisfied.

If $c < \aleph_0$, (ii) is trivially met. Thus, let $c = \aleph_0$ and choose $T_0 \in R^1$ and $x \in H^c$. If $\varepsilon > 0$, find $N > 0$ so that $\sum_{k=N+1}^{\infty} \|x_k\|^2 < \frac{\varepsilon^2}{8}$; by our hypothesis on E_T, there exists intervals $I_i = [T_0, T_i)$, $i = 1,2,\ldots,N$ such that $\|E_T x_i - E_{T_0} x_i\|^2 < \frac{\varepsilon^2}{2N}$ for each $T \in I_i$ and $i = 1,2,\ldots,N$. Putting $I = \bigcap_{i=1}^{N} I_i$, we have for $T \in I$, $\|S_T x - S_{T_0} x\|_c^2 = \sum_{i=1}^{\infty} \|E_T x_i - E_{T_0} x_i\|^2 = \sum_{i=1}^{N} \|\ldots\|^2 + \sum_{i=N+1}^{\infty} \|\ldots\|^2 < N \cdot \frac{\varepsilon^2}{2N} + \sum_{i=N+1}^{\infty} (\|E_T\| + \|E_{T_0}\|)^2 \|x_i\|^2 < \varepsilon^2$. Hence, $S_T x \to S_{T_0} x$ in H^c as $T \to T_0$, $T > T_0$.

The verification of (iii) is similar.

At this point, let us discuss some specific cases of spaces with a resolution of the identity.

Example 1. Let $H = R^{\aleph_o} = \ell_2$, and for each $x = [x_k] \in \ell_2$ and $T \in R^1$, let $E_T = [x_1, x_2, \ldots, x_m, 0, 0, \ldots]^T$, where m is the largest index such that $m \leq T$. We can easily verify that $\{E_T : T \in R^1\}$ is a resolution of identity on ℓ_2.

Example 2. Let $H = L_2(-\infty, \infty)$; for every $x \in L_2(-\infty, \infty)$ and $T \in R^1$, let

$$(E_T x)(t) = x(t) \text{ for } t \in (-\infty, T], \qquad (2.6.2)$$
$$= 0 \text{ for } t \in (T, \infty).$$

Again, we can easily show that $\{E_T : T \in R^1\}$ is a resolution of identity on $L^2(-\infty, \infty)$. Note that the element $E_T x$ is sometimes called the "truncation of x at T".

Example 3. Let $H = L_2[o, \tau]$, $\tau > 0$, and let η be a continuous increasing function on R^1 which maps R^1 onto (o, τ). For every $T \in R^1$ and $x \in L_2[o, \tau]$, let

$$(E_T x)(t) = x(t) \text{ for } t \in [o, \eta(T)], \qquad (2.6.3)$$
$$= 0 \text{ for } t \in (\eta(T), \tau].$$

Then $\{E_T : T \in R^1\}$ is a resolution of identity on $L_2[o, \tau]$.

Observe this important fact: If $c \leq \aleph_o$ and $M : L_2^c[o, \tau] \to \mathfrak{S}(L_2^c[o, \tau])$ is a set mapping, then causality of M on $L_2^c[o, \tau]$, with E_T being defined by (2.6.3) and S_T by (2.6.1), clearly means that we have the implication $x_1(t) = x_2(t)$ on $[o, T']$, $0 < T' < \tau \Rightarrow (Mx_1)(t) = (Mx_2)(t)$ on $[o, T']$. Thus if t is interpreted as time and M describes some physical system, then this system is unanticipative.

Note that a similar comment holds for examples 1 and 2.

Lemma 2.6.2. Let $\{E_T : T \in R^1\}$ be a resolution of identity on H, let $S_T : H^{c_2} \to H^{c_2}$ and $S_T' : H^{c_o} \to H^{c_o}$ be defined by (2.6.1) and let $\hat{X} \in [H^{c_o}, H^{c_2}]$ have the usual meaning. Then

$$\hat{X} S_T' = S_T \hat{X}, \quad S_T' \hat{X}^* = \hat{X}^* S_T \qquad (2.6.4)$$

for every $T \in R^1$, and the collection $\{S_T : T \in R^1\}$ satisfies the condition (i) in Theorem 1.6.1, i.e., for each $T \in R^1$ the projection S_T commutes with $P = \hat{X}\hat{X}^*$.

Proof: Let $X = [\xi_{ik}]$ be the $c_2 \times c_o$ matrix generating the operator $\hat{X}$, and let $u = [u_k] \in H^{c_o}$; then we have by linearity and continuity of E_T, $\hat{X}S'_T u = [\xi_{ik}]\cdot[E_T u_k] = [\sum_k \xi_{ik}(E_T u_k)] = [E_T(\sum_k \xi_{ik} u_k)] = S_T\hat{X}u$. Hence, the first equality (2.6.4) holds; the second one follows by taking adjoints. Moreover, we have $PS_T = \hat{X}\hat{X}^*S_T = \hat{X}S'_T\hat{X}^* = S_T\hat{X}\hat{X}^* = S_TP$ what we wanted to show.

<u>Theorem 2.6.1</u>. Let $\{E_T : T \in R^1\}$ be a resolution of identity on H, and let S_T, S'_T be defined by (2.6.1); moreover, let $\hat{n} = (\hat{Z}, G)$ be a Hilbert network, and let $W = \hat{X}^*\hat{Z}\hat{X} : F \to \mathfrak{S}(WF)^\circ)$. Assume that

$$\langle w_1 - w_2, y_1 - y_2\rangle_{c_o} \neq 0 \tag{2.6.5}$$

for all $y_1, y_2 \in F$, $y_1 \neq y_2$ and all $w_1 \in Wy_1$, $w_2 \in Wy_2$, and that

(i)* $S'_T F \subset F$ for each $T \in R^1$,

(ii)* the set mapping W is causal on F with respect to $\{S'_T : T \in R^1\}$.

Then the admittance A of $\hat{n}$ is causal on Q(D) with respect to $\{S_T : T \in R^1\}$.

<u>Proof</u>: By Lemma 2.6.2, $S'_T F \subset F \Leftrightarrow \hat{X}S'_T F \subset \hat{X}F \Leftrightarrow S_T\hat{X}F \subset \hat{X}F$, i.e., $S_T(N_{\hat{a}} \cap D) \subset N_{\hat{a}} \cap D$; thus (i)* implies condition (ii) in Theorem 1.6.1.

Moreover, we have by Lemma 1.6.1, W is causal on F $\Leftrightarrow S'_T W = S'_T W S'_T$ on F for all $T \in R^1 \Leftrightarrow S'_T\hat{X}^*\hat{Z}\hat{X} = S'_T\hat{X}^*\hat{Z}S_T\hat{X}$ on F for all $T \in R^1 \Leftrightarrow S'_T\hat{X}^*\hat{Z} = S'_T\hat{X}^*\hat{Z}S_T$ on $N_{\hat{a}} \cap D$ for all $T \in R^1 \Leftrightarrow \hat{X}S'_T\hat{X}^*\hat{Z} = \hat{X}S'_T\hat{X}^*\hat{Z}S_T$ on $N_{\hat{a}} \cap D$ for all $T \in R^1 \Leftrightarrow S_TP\hat{Z} = S_TP\hat{Z}S_T$ on $N_{\hat{a}} \cap D$ for all $t \in R^1 \Leftrightarrow P\hat{Z}$ is causal on $N_{\hat{a}} \cap D$; hence, (ii)* implies (iii) in Theorem 1.6.1. The rest of the proof is obvious.

<u>Theorem 2.6.2</u>. Let $\{E_T : T \in R^1\}$ be a resolution of identity on H, and let S'_T be defined by (2.6.1); furthermore, let $\hat{n} = (\hat{Z}, G)$ be a Hilbert network and let $W = \hat{X}^*\hat{Z}\hat{X} : F \to WF$. Assume that there exists constants $c > 0$ and $p > 1$ such that

$$|\langle w_1 - w_2, S'_T(y_1 - y_2)\rangle_{c_o}| \geq c\|S'_T(y_1 - y_2)\|^p_{c_o} \tag{2.6.6}$$

for all $T \in R^1$, $y_j \in F$ and $w_j \in Wy_j$, $j = 1,2$. Then $\hat{n}$ is regular on D and the admittance A of $\hat{n}$ is causal on Q(D), (with respect to the family $\{S_T : T \in R^1\}$.)

The proof of Theorem 2.6.2 is obvious and therefore omitted.

Note that Theorem 1.6.2 remains valid without change for Hilbert networks; also, referring to the Remark following Theorem 1.6.3, we see that we have:

<u>Proposition</u>: The condition (2.6.6) in Theorem 2.6.2 may be replaced by the stronger requirement that $|\langle z_1-z_2, S_T(x_1-x_2)\rangle_{c_2}| \geq c\|S_T(x_1-x_2)\|_{c_2}^p$ for all $T \in R^1$, $x_j \in D$ and $z_j \in \hat{Z}x_j$, $j = 1,2$.

CHAPTER 3

SOME EXAMPLES OF SPECIFIC HILBERT NETWORKS

In this part we will discuss several networks when the underlying space H is a specific function space, or possibly R^1.

3.1 An Auxiliary Result

In order to investigate infinite networks, let us introduce the following concept:

A $c \times c$ matrix $M = [m_{ik}]$, $c \le \aleph_o$, will be called a ribbon matrix, if there exists an integer $p \ge 1$ such that $m_{ik} = 0$ whenever $|i-k| \ge p$.

Lemma 3.1.1. Let H, $\tilde{H}$ be Hilbert spaces and let $Y = [Y_{ik}]$ be a $c \times c$ matrix ($c \le \aleph_o$) whose elements Y_{ik} are operators from H into $\tilde{H}$. If

(a) Y is a ribbon matrix,

(b) there exists an $a > 0$ such that $Y_{ik} \in Lip_a[H,\tilde{H}]$ for all i,k,

(c) $Y_{ik}0 = 0$ for all i,k,

then the operator $\hat{Y}$ defined on H^c by $\hat{Y}x = [Y_{ik}]\cdot x$ is in $Lip_{a'}[H^c,\tilde{H}^c]$ with $a' = (2p+1)a$. If in addition, $c < \aleph_o$, the assumption (c) can be dropped.

Proof: First, observe that due to (a), for any $x = [x_k] \in H^c$, the sum $\sum_\mu Y_{i\mu}x_\mu$ involves at most 2p+1 nonzero terms for any i; thus, the element $\sum_\mu Y_{i\mu}x_\mu$ is in $\tilde{H}$.

Next, choose $u,v \in H^c$, and denote $\hat{Y}u = [y_k]$, $\hat{Y}v = [z_k]$; then we have by Schwartz inequality and (b),

$$\sum_{i=1}^m \|y_i - z_i\|^2 = \sum_{i=1}^m \|\sum_\mu Y_{i\mu}u_\mu - Y_{i\mu}v_\mu\|^2 \qquad (1.3.1)$$

$$\le \sum_{i=1}^m \Big(\sum_{\substack{|\mu-i|\le p \\ \mu\ge 1}} a\|u_\mu - v_\mu\|\Big)^2 \le \sum_{i=1}^m a^2(2p+1) \sum_{\substack{|\mu-i|\le p \\ \mu\ge 1}} \|u_\mu - v_\mu\|^2$$

$$\le a^2(2p+1)^2\|u-v\|_c^2.$$

Putting v = 0 into (3.1.1) for a while, we get by (c), $\sum_{i=1}^{m} \|y_i\|^2 \leq a^2(2p+1)^2\|u\|_c^2$, i.e., $[y_k] \in \tilde{H}^c$; consequently, $\hat{Y}$ maps H^c into $\tilde{H}^c$. Moreover, (3.1.1) implies that $\|y-z\|_c^2 \leq a^2(2p+1)^2\|u-v\|_c^2$, i.e., $\hat{Y} \in \mathrm{Lip}_{a'}[H^c,\tilde{H}^c]$ which we wanted to show.

Finally, let us introduce the following terminology: If M is a real constant c × c matrix, ($c \leq \aleph_0$), we will say that M is positive semidefinite, if for every real $\xi \in R^c$ we have $M\cdot\xi \in R^c$ and $\xi^T\cdot(M\cdot\xi) \geq 0$.

3.2 A Hilbert Network with $H = L_2[0,T]$

(A) In this example we will assume that H is the real Hilbert space $L_2[0,T]$, $0 < T < \infty$.

Let $\mathfrak{G}$ be a locally finite oriented graph having the set of branches $\mathfrak{B}$, and let $c_2 \leq \aleph_0$ be the cardinal of $\mathfrak{B}$. Furthermore, let us make the following assumptions:

(a) For every index j, let r_j be a set mapping from R^1 (real line) into $\mathfrak{S}(R^1)$; assume that there exists a constant $\alpha > 0$ such that, for any $\sigma_1,\sigma_2 \in R^1$ and any $\omega_1 \in r_j(\sigma_1)$, $\omega_2 \in r_j(\sigma_2)$ we have

$$\alpha(\sigma_1-\sigma_2)^2 \leq (\omega_1-\omega_2)(\sigma_1-\sigma_2). \qquad (3.2.1)$$

Moreover, let there exist an integer $N > 0$ and a constant $\beta > 0$ such that, for all $j > N$, $\sigma \in R^1$ and $\omega \in r_j(\sigma)$ we have

$$|\omega| \leq \beta|\sigma|. \qquad (3.2.2)$$

(b) For every pair of indices i,k let $L_{ik}(t)$ be a real function having a derivative everywhere on $[0,T]$. Assume that there exists $\gamma > 0$ such that

$$|L_{ik}(t)| \leq \gamma, \quad |L'_{ik}(t)| \leq \gamma \qquad (3.2.3)$$

for all $t \in [0,T]$ and all i,k. Moreover for each $t \in [0,T]$, let $[L_{ik}(t)]$ and $[L'_{ik}(t)]$ be ribbon, symmetric and positive semidefinite $c_2 \times c_2$ matrices.

(c) For every pair of indices i,k, let $S_{ik}(t)$ be a real function having a bounded derivative on $[0,T]$, and let a $\delta > 0$ exist such that

$$|S_{ik}(t)| \leq \delta \qquad (3.2.4)$$

for all $t \in [0,T]$. Assume that, for each $t \in [0,T]$, $[S_{ik}(t)]$ and $-[S'_{ik}(t)]$ are ribbon, symmetric and positive semidefinite matrices on $[0,T]$.

Finally, let $i_0 \in R^{c_2}$ be a vector satisfying the equation $a\cdot i_0 = 0$ with

a being the $c_1 \times c_2$ matrix constructed in Section 2.1 from the incidence matrix d of G.

Let $D \subset L_2^{c_2}[o,\tau]$ be the set of all c_2-vectors $x = [x_k(t)]$ such that each $x_k(t)$ is absolutely continuous on $[o,\tau]$, $x' = [x_k'(t)] \in L_2^{c_2}[o,\tau]$ and $x(0) = i_o$.

Let the set mapping $\hat{Z}$ be defined on D by

$$\hat{Z}y = [Z_{ik}]\cdot[y_k], \tag{3.2.5}$$

where each set mapping Z_{ik} is defined by

$$(Z_{ik}\xi)(t) = (L_{ik}(t)\xi)' + \eta_{ik}(t) + S_{ik}(t)\int_o^t \xi(\sigma)d\sigma; \tag{3.2.6}$$

here, $\eta_{ik}(t)$ is any measurable function such that $\eta_{ik}(t) \in r_{ik}(\xi(t))$ for each $t \in [o,\tau]$. Also, we let $r_{ik} = 0$ for $i \neq k$ and $r_{ii} = r_i$, $i = 1,2,3,\ldots$

Let us consider the network $\hat{n} = (\hat{Z},G)$; we are going to show that, under the assumptions made, $\hat{n}$ is regular on D, the admittance A of $\hat{n}$ is in $\mathrm{Lip}_{\alpha^{-1}}[Q(D),L_2^{c2}[o,\tau]]$, and A is causal on $Q(D)$ with respect to the resolution of identity $\{S_T : T \in R^1\}$ discussed in Example 3, Section 2.6. Thus, we consider an L, R, C-network in the time domain, which consists of time-varying inductors and capacitors, and of time-invariant nonlinear multi-valued resistors. Also, we are assuming that the initial current regime $(t = 0)$ in the network is described the the vector i_o. Note that, due to condition (3.2.1), the characteristic r_j of a resistor may appear as is indicated in Figure 3.2.

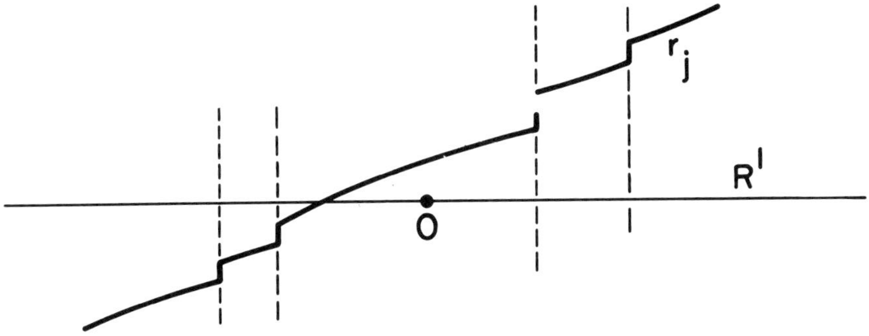

Figure 3.2

To prove our assertion, define operators V_{ik}, U_{ik} and the set mappings R_{ik} by

$$(V_{ik}\xi)(t) = (L_{ik}(t)\xi)', \quad (U_{ik}\xi)(t) = S_{ik}(t)\int_0^t \xi(\sigma)d\sigma, \qquad (3.2.7)$$

$$(R_{ik}\xi)(t) = r_{ik}(\xi(t)), \quad t \in [o,\tau];$$

also, let the operators $\hat{V}$, $\hat{U}$ and the set mapping $\hat{R}$ be defined on D by

$$\hat{V}y = [V_{ik}]\cdot[y_k], \quad \hat{U}y = [U_{ik}]\cdot[y_k], \qquad (3.2.8)$$

$$\hat{R}y = [R_{ik}]\cdot[y_k].$$

Then clearly $\hat{Z} = \hat{V} + \hat{R} + \hat{U}$.

Since $\hat{V}y = \{[L_{ik}(t)]\cdot[y_k]\}' = [L'_{ik}(t)]\cdot[y_k] + [L_{ik}(t)]\cdot[y'_k]$, it follows from (3.2.3) and the fact that $[L_{ik}(t)]$ and $[L'_{ik}(t)]$ are ribbon matrices, that $\hat{V}y \in L_2^{c_2}[o,\tau]$ for each $y \in D$, (see Lemma 3.1.1).

Now let $T \in R^1$ and denote $\tau' = \eta(T)$, where η is the function introduced in Example 3, Section 2.6; then $0 < \tau' < \tau$. If $x,y \in L_2^{c_2}[o,\tau]$, $x = [x_k]$, $y = [y_k]$, we get easily by a standard theorem (see Bartle, [9], p. 50),

$$\begin{aligned}\langle x,S_Ty\rangle_{c_2} &= \sum_k \langle x_k,E_Ty_k\rangle = \sum_k \int_0^\tau x_k(t)(E_Ty_k)(t)dt \qquad (3.2.9)\\ &= \sum_k \int_0^{\tau'} x_k(t)y_k(t)dt = \int_0^{\tau'} \sum_k x_k(t)y_k(t)dt = \\ &= \int_0^{\tau'} x^T(t)\cdot y(t)dt,\end{aligned}$$

where $x^T(t)\cdot y(t)$ converges almost everywhere in $[o,\tau]$. Thus, if $x_1,x_2 \in D$, we can write with $L = [L_{ik}]$, $J = \langle \hat{V}x_1-\hat{V}x_2,S_T(x_1-x_2)\rangle_{c_2} = \int_0^{\tau'} (x_1-x_2)^T\cdot\{L\cdot(x_1-x_2)\}'d\sigma = [(x_1-x_2)^T\cdot(L\cdot(x_1-x_2)]_0^{\tau'} - \int_0^{\tau'}(x'_1-x'_2)\cdot(L\cdot(x_1-x_2))d\sigma$. Also $J = \int_0^{\tau'} (x_1-x_2)^T\cdot(L'\cdot(x_1-x_2))d\sigma + \int_0^{\tau'} (x_1-x_2)^T\cdot(L\cdot(x'_1-x'_2))d\sigma$. Since $x_1(0) = x_2(0) = i_o$, we get from these equalities due to symmetry of L,

$$\begin{aligned}J &= \tfrac{1}{2}(x_1-x_2)^T(\tau')\cdot(L(\tau')\cdot(x_1-x_2)(\tau)) \qquad (3.2.10)\\ &+ \tfrac{1}{2}\int_0^{\tau'} (x_1-x_2)^T(\sigma)\cdot(L'(\sigma)\cdot(x_1-x_2)(\sigma))d\sigma.\end{aligned}$$

Hence, by positive semi-definiteness of L and L', $J \geq 0$.

Similarly, (3.2.2) shows easily that $\hat{R}x \subset L_2^{c_2}[o,\tau]$ whenever $x \in D$. Also, a routine calculation confirms that (3.2.1) implies that

$$\langle w_1-w_2,S_T(x_1-x_2)\rangle_{c_2} \geq \alpha\|S_T(x_1-x_2)\|_{c_2}^2 \qquad (3.2.11)$$

whenever $x_1,x_2 \in D$ and $w_1 \in \hat{R}x_1$, $w_2 \in \hat{R}x_2$.

Next, let $x_j \in D$, $j = 1,2$, and denote $y_j(t) = \int_0^t x_j(\sigma)d\sigma$; it is clear that $y_j \in H^{c_2}$, and we have by (3.2.7), (3.2.8) and (3.2.9) with $S = [S_{ik}]$,

$$J' = \langle \hat{U}x_1 - \hat{U}x_2, S_T(x_1-x_2)\rangle_{c_2}$$
$$= \int_0^{\tau'} (x_1-x_2)^T(\sigma)\cdot(S(\sigma)\cdot(\int_0^{\sigma} x_1(\omega)d\omega - \int_0^{\sigma} x_2(\omega)d\omega))d\sigma$$
$$= [(y_1-y_2)^T\cdot(S\cdot(y_1-y_2))]_0^{\tau'} - \int_0^{\tau'} (y_1-y_2)^T\cdot(S\cdot(y_1-y_2))'d\sigma$$
$$= [(y_1-y_2)^T\cdot(S\cdot(y_1-y_2))]_0^{\tau'} - \int_0^{\tau'} (y_1-y_2)^T\cdot(S'\cdot(y_1-y_2))d\sigma$$
$$- \int_0^{\tau'} (y_1-y_2)^T\cdot(S\cdot(x_1-x_2))d\sigma.$$

Using the symmetry of S' and the fact that $y_1(0) = y_2(0) = 0$, we get from the above equality,

$$J' = (y_1-y_2)^T(\tau')\cdot(S(\tau')\cdot(y_1-y_2)(\tau')) - \int_0^{\tau'} (y_1-y_2)^T$$
$$\cdot(S'\cdot(y_1-y_2))d\sigma - J';$$

consequently,

$$J' = \tfrac{1}{2}(y_1-y_2)^T(\tau')\cdot(S(\tau')\cdot(y_1-y_2)(\tau') - \int_0^{\tau'} (y_1-y_2)^T$$
$$\cdot(S'\cdot(y_1-y_2))d\sigma.$$

Thus, by our hypothesis (c), $J' \geq 0$.

Combining these results with the definition of $\hat{Z}$, it follows that

$$\langle z_1-z_2, S_T(x_1-x_2)\rangle_{c_2} \geq \alpha\|S_T(x_1-x_2)\|^2_{c_2} \qquad (3.2.12)$$

for all $x_1, x_2 \in D$ and all $z_1 \in \hat{Z}x_1$, $z_2 \in \hat{Z}x_2$. Hence, Theorem 2.6.2 and the Proposition following it conclude the proof of our assertion. (Note that $A \in \mathrm{Lip}_{\alpha^{-1}}[Q(D), L_2^{c_2}[o,\tau]]$ is guaranteed by Theorem 1.2.5 (c), since, as we have seen before, (3.2.12) implies (1.2.10) by letting $T \to \infty$; also, since $\|P\| = 1$, $\|P(e_1-e_2)\|_{c_2} \leq \|e_1-e_2\|_{c_2}$ for all $e_1, e_2 \in Q(D)$.)

Remark 1. Using physical terminology, the above assumption (a) means that each branch of our network contains a nonzero, nonlinear, multivalued resistor. However, from our development it is clear that (a) can be replaced by the weaker requirement that, with $\tilde{R} = \hat{X}^*R\hat{X}$, we have

$$\langle z_1-z_2, S_T(y_1-y_2)\rangle_{c_o} \geq \alpha\|S_T(y_1-y_2)\|^2_{c_o}$$

for all $y_1, y_2 \in H^{c_o}$ and $z_1 \in \tilde{R}y_1$, $z_2 \in \tilde{R}y_2$. This assumption allows the situation when some branches do not contain any resistor, but, of course, "each loop contains some resistor".

(B) Let us now modify the previous example as follows: First, assume that $L_{ik}(t) = 0$ for every i,k, i.e., our network does not contain any inductors.

Moreover, assume that each r_j is a function on R^1 satisfying the conditions (3.2.1), (3.2.2) such that the operator $\hat{R}$, defined by (3.2.8), is continuous. (Referring to Lemma 3.1.1, this condition will be satisfied, if $r_j \in Lip_{\alpha'}[R^1,R^1]$ for every j, $\alpha' > 0$ fixed, and $r_j(0) = 0$.)

As for the S_{ik}'s, let the conditions given in (c) hold.

In this case, let $D = L_2^{c_2}[o,\tau]$. (Clearly, since the differentiators are not present in the network, operators $\hat{U}$, $\hat{R}$ are well-defined on the entire space $L_2^{c_2}[o,\tau]$.)

Using Lemma 3.1.1 and (3.2.4), we conclude easily that the operator $\hat{U}$ satisfies the Lipschitz condition on $L_2^{c_2}[o,\tau]$; consequently, the operator $\hat{Z} = \hat{R} + \hat{U}$ is continuous on $L_2^{c_2}[o,\tau]$. Furthermore, (3.2.12) implies that we have

$$\langle \hat{Z}x_1-\hat{Z}x_2,x_1-x_2\rangle_{c_2} \geq \alpha\|x_1-x_2\|_{c_2}^2 \tag{3.2.13}$$

for all $x_1,x_2 \in L_2^{c_2}[o,\tau]$.

Hence, by Theorem 1.3.2, $\hat{\eta}$ is regular on $L_2^{c_2}[o,\tau]$ and $Q(L_2^{c_2}[o,\tau]) = L_2^{c_2}[o,\tau]$, i.e., for any $e \in L_2^{c_2}[o,\tau]$, $\hat{\eta}$ possesses a unique solution corresponding to e.

3.3 Periodic Regime in an R,C-Network.

Let H be the set of all real-valued functions on R^1 (real line), which are periodic with period 2τ, $\tau > 0$, measurable and such that $\int_{-\tau}^{\tau} |x(t)|^2 dt < \infty$ whenever $x \in H$. It is clear that H becomes a separable Hilbert space, if we put

$$\langle x,y\rangle = \int_{-\tau}^{\tau} x(t)y(t)dt. \tag{3.3.1}$$

Moreover, let

$$\tilde{H} = \{x : x \in H, \int_{-\tau}^{\tau} x(t)dt = 0\}. \tag{3.3.2}$$

It is easy to see that $\tilde{H}$ is a closed linear subspace of H.

Next, let G be a locally finite oriented graph having the set of branches $\mathfrak{B}$, and let $c_2 \leq \aleph_0$ be the cardinal of $\mathfrak{B}$. Furthermore, let the following assumptions be satisfied:

(a) For every index j, let r_j be a real-valued function on R^1; assume that there exists a constant $\alpha > 0$ such that, for any $\sigma_1, \sigma_2 \in R^1$,

$$\alpha(\sigma_1 - \sigma_2)^2 \leq (r_j(\sigma_1) - r_j(\sigma_2))(\sigma_1 - \sigma_2). \tag{3.3.3}$$

Also, let an integer $N > 0$ and a constant $\beta > 0$ exist such that

$$|r_j(\sigma)| \leq \beta|\sigma| \tag{3.3.4}$$

for all $j > N$, $\sigma \in R^1$.

(b) Let S_j be real numbers, and let

$$|S_j| \leq \delta \tag{3.3.5}$$

for all j and some $\delta > 0$. Let the $c_2 \times c_2$ matrix S be defined by

$$S = \mathrm{diag}(S_1, S_2, S_3, \ldots). \tag{3.3.6}$$

Finally, let $D = \tilde{H}^{c_2} \subset H^{c_2}$, and let the operator $\hat{Z}$ be defined on D by

$$\hat{Z}y = \mathrm{diag}(Z_1, Z_2, Z_3, \ldots)\cdot[y_k], \tag{3.3.7}$$

when each Z_j is defined by

$$(Z_j\xi)(t) = r_j(\xi(t)) + S_j \int_0^t \xi(\sigma)d\sigma. \tag{3.3.8}$$

Consider the network $\hat{n} = (\hat{Z}, G)$; we will show that, with the above assumptions, $\hat{n}$ is regular on D. Using physical terminology, we consider an R, C network in the time domain, whose capacitors are constant, and whose resistors are nonlinear and time invariant. Then the regularity means that any 2π-periodic solution with average value zero of our network driven by 2π-periodic EMF's is determined uniquely.

To prove our assertion, observe first that from (3.3.3), (3.3.4) it follows in the same way as in Section 3.2 that the operator $\hat{R}$, defined on D by

$$\hat{R}[y_k] = [r_1(y_1(t)), r_2(y_2(t)), \ldots]^T, \tag{3.3.9}$$

maps D into H^{c_2}, and that

$$\langle \hat{R}x_1 - \hat{R}x_2, x_1 - x_2 \rangle_{c_2} \geq \alpha \|x_1 - x_2\|_{c_2}^2 \tag{3.3.10}$$

for all $x_1, x_2 \in D$.

On the other hand, if $\xi \in \tilde{H}$, we see readily that $\eta(t) = \int_0^t \xi(\sigma)d\sigma$ is in H; indeed, $\eta(t+2\tau) - \eta(t) = \int_0^{t+2\tau}\xi(\sigma)d\sigma - \int_0^t \xi(\sigma)d\sigma = \int_t^{t+2\tau}\xi(\sigma)d\sigma$ $= \int_{-\tau}^{\tau} \xi(\sigma)d\sigma = 0$, i.e., $\eta(t)$ is 2τ-periodic. Thus, using Lemma 3.1.1, it follows that the operator $\hat{U}$ defined on D by

$$\hat{U}y = [S_k \int_0^t y_k(\sigma)d\sigma] \tag{3.3.11}$$

maps D into H^{c_2} (witness condition 3.3.5)). Moreover, if $y \in D$ and $z = \int_0^t y(\sigma)d\sigma$, we get $M = \langle \hat{U}y,y\rangle_{c_2} = \int_{-\tau}^{\tau} y^T(t)\cdot[S\cdot z(t)]dt = \{z^T(t)\cdot(S\cdot z(t)\}_{-\tau}^{\tau}$ $- \int_{-\tau}^{\tau} z^T(t)\cdot[S\cdot y(t)]dt$. Hence, by symmetry of S, $M = \frac{1}{2}\{z^T(t)\cdot(S\cdot z(t))\}_{-\tau}^{\tau}$. However, since z is 2τ-periodic, we have $z(-\tau) = z(\tau)$, and consequently, $M = 0$. Thus, combining this with (3.3.10) and (3.3.8), it follows that

$$\langle \hat{Z}y_1 - \hat{Z}y_2, y_1 - y_2\rangle_{c_2} \geq \alpha\|y_1 - y_2\|_{c_2}^2 \tag{3.3.12}$$

for all $y_1, y_2 \in D$. (Note that the operator $\hat{U}$ is linear.) Consequently, $\hat{h}$ is regular on D by Theorem 1.2.5, what we wanted to show.

3.4 L, R, C-Networks with Constant Elements in the Frequency Domain

In this section we will consider in more detail finite or infinite L, R, C-networks with constant elements driven by voltage sources e_j of the form $e_j = \alpha_j \cos \omega t + \beta_j \sin \omega t$. This case is most common in classical network analysis, and is currently called "the frequence domain analysis" [6].

First, we will show that the problem can be described by a network whose underlying space if an Euclidean space R^{c_2}, and then, using this fact, we will give necessary and sufficient conditions for the existence and uniqueness of a regime in an L, R, C-network.

Since we will have to distinguish between the real and complex spaces R^c, $c \leq \aleph_0$, we will make the following agreement: R^1 will stand for the real line, and $\bar{R}^1$ for the complex plane. The meanings of R^c and $\bar{R}^c$ will be analogous.

Let $\omega > 0$ be a fixed real number, and let

$$V_\omega = \{\alpha \cos \omega t + \beta \sin \omega t : \alpha, \beta \in R^1\}; \tag{3.4.1}$$

clearly, V_ω is a real linear space. If, for $x_j = \alpha_j \cos \omega t + \beta_j \sin \omega t$, $j = 1,2$, we put

$$\langle x_1, x_2\rangle = \alpha_1\alpha_2 + \beta_1\beta_2, \tag{3.4.2}$$

then $\langle\cdot,\cdot\rangle$ is an inner product on V_ω, and V_ω becomes a separable Hilbert space. Also note that there is an isomorphism between V_ω and $\bar{R}^1$ described by $V_\omega \ni x \leftrightarrow \alpha - i\beta \in \bar{R}^1$.

Next, let $c \leq \aleph_0$, and define V_ω^c by (2.1.1). It is clear that $\alpha \cos \omega t + \beta \sin \omega t \in V_\omega^c \Leftrightarrow \alpha, \beta \in R^c$. Also, if $x_j = \alpha_j \cos \omega t + \beta \sin \omega t \in V_\omega^c$, $j = 1,2$, then

$$\langle x_1, x_2\rangle_c = \sum_k \langle \alpha_k^1 \cos \omega t + \beta_k^1 \sin \omega t, \alpha_k^2 \cos \omega t + \beta_k^2 \sin \omega t\rangle \quad (3.4.3)$$
$$= \sum_k \alpha_k^1\alpha_k^2 + \beta_k^1\beta_k^2 = \alpha_1^T\cdot\alpha_2 + \beta_1^T\cdot\beta_2.$$

Furthermore, define a transformation $\mathfrak{J} : V_\omega^c \to \bar{R}^c$ by

$$\mathfrak{J}(\alpha \cos \omega t + \beta \sin \omega t) = \alpha - i\beta; \quad (3.4.4)$$

clearly, $\mathfrak{J}$ is an isomorphism between V_ω^c and $\bar{R}^c$ ($\bar{R}^c$ being considered as a real space). Moreover, if $\bar{R}^c$ is equipped with the standard inner product, i.e., if $\langle\xi_1, \xi_2\rangle = \xi_1^T\cdot\bar{\xi}_2$ for $\xi_1, \xi_2 \in \bar{R}^c$, then

$$\langle x_1, x_2\rangle_c = \mathrm{Re}\,\langle\mathfrak{J}x_1, \mathfrak{J}x_2\rangle \quad (3.4.5)$$

for all $x_j \in V_\omega^c$, $j = 1,2$. Indeed, if $x_j = \alpha_j \cos \omega t + \beta_j \sin \omega t$, $j = 1,2$, then $\mathfrak{J}x_j = \alpha_j - i\beta_j \in \bar{R}^c$, and $\langle\mathfrak{J}x_1, \mathfrak{J}x_2\rangle = (\alpha_1 - i\beta_1)^T\cdot\overline{(\alpha_2 - i\beta_2)} = (\alpha_1 - i\beta_1)^T\cdot(\alpha_2 + i\beta_2)$ $= \alpha_1^T\cdot\alpha_2 + \beta_1^T\cdot\beta_2 + i(\alpha_1^T\cdot\beta_2 - \beta_1^T\cdot\alpha_2)$; by (3.4.3), however, $\langle x_1, x_2\rangle_c = \alpha_1^T\cdot\alpha_2 + \beta_1^T\cdot\beta_2$, which confirms (3.4.5). Also, observe that (3.4.5) yields for $x \in V_\omega^c$,

$$\|x\|_c = \|\mathfrak{J}x\|, \quad (3.4.6)$$

i.e., $\mathfrak{J}$ is a norm preserving isomorphism.

Now, let G be a locally finite oriented graph, and let $c_2 \leq \aleph_0$ be the cardinal of the set of branches in G; let L, R, S be real $c_2 \times c_2$ matrices such that $L\cdot\gamma$, $R\cdot\gamma$, $S\cdot\gamma \in R^{c_2}$ for every $\gamma \in R^{c_2}$. Furthermore, let the operator $\hat{Z} : V_\omega^{c_2} \to V_\omega^{c_2}$ be defined by

$$\hat{Z}x = L\cdot x' + R\cdot x + S\circ\textstyle\int x; \quad (3.4.7)$$

here, $x' = -\omega\alpha \sin \omega t + \omega\beta \cos \omega t$ and $\int x = \omega^{-1}\alpha \sin \omega t - \omega^{-1}\beta \cos \omega t$ for each $x = \alpha \cos \omega t + \beta \sin \omega t \in V_\omega^{c_2}$. The operator $\hat{Z}$ truly maps $V_\omega^{c_2}$ into $V_\omega^{c_2}$. Indeed, if $\alpha = \alpha \cos \omega t + \beta \sin \omega t \in V_\omega^{c_2}$, then

$$\hat{Z}x = \{\omega L\cdot\beta + R\cdot\alpha - \omega^{-1}S\cdot\beta\}\cos \omega t \quad (3.4.8)$$
$$+ \{-\omega L\cdot\alpha + R\cdot\beta + \omega^{-1}S\cdot\alpha\}\sin \omega t;$$

however, since both the vectors $\{\ldots\}$ are in R^{c_2}, we have $\hat{Z}x \in V_\omega^{c_2}$.

Next, let the $c_2 \times c_2$ matrix M_ω be defined by

$$M_\omega = i\omega L + R - i\omega^{-1}S, \tag{3.4.9}$$

and let $Z^+ : \bar{R}^{c_2} \to \bar{R}^{c_2}$ be defined by

$$Z^+\xi = M_\omega \cdot \xi. \tag{3.4.10}$$

Then we have the assertion:

If $x \in V_\omega^{c_2}$, then $\mathfrak{J}(\hat{Z}x) = Z^+(\mathfrak{J}x)$.

Indeed, by (3.4.8), $\mathfrak{J}(\hat{Z}x) = \{\omega L\cdot\beta + R\cdot\alpha - \omega^{-1}S\cdot\beta\} - i\{-\omega L\cdot\alpha + R\cdot\beta + \omega^{-1}S\cdot\alpha\} = M_\omega\cdot(\alpha - i\beta) = Z^+(\mathfrak{J}x)$.

Now, let $\hat{\mathfrak{n}} = (\hat{Z},G)$ with the underlying space being V_ω, and let $\mathfrak{n}^+ = (Z^+,G)$ with the underlying space $\bar{R}^1$.

Before proceeding further, let us explain the physical meaning of these concepts.

Suppose that $e \in V_\omega^{c_2}$, and that $\tilde{j} \in V_\omega^{c_2}$ has the meaning of a vector of currents in $\hat{\mathfrak{n}}$ forced by e. Then, by (3.4.7), the voltage drop in a branch b_j is given by $\sum_k(L_{jk}\tilde{j}'_k + R_{jk}\tilde{j}_k + S_{jk}\int\tilde{j}_k)$, i.e., b_j contains mutual inductaces, resistances and capacities given by numbers $L_{jk}, R_{jk}, S_{jk}^{-1}$.

Similarly, if $e^+ \in \bar{R}^{c_2}$ and $j^+ \in \bar{R}^{c_2}$ is the vector of currents in $\mathfrak{n}^+$ corresponding to e^+, then the voltage drop in a branch b_j is given by $\sum_k(i\omega L_{jk} + R_{jk} + (i\omega)^{-1}S_{jk})j_k^+$.

From the above considerations it follows easily that the following claim is true:

<u>Proposition 1.</u> Let $e \in V_\omega^{c_2}$; then $j \in V_\omega^{c_2}$ is a solution of $\hat{\mathfrak{n}}$ corresponding to e, if and only if $\mathfrak{J}j \in \bar{R}^{c_2}$ is a solution of $\mathfrak{n}^+$ corresponding to $\mathfrak{J}e \in \bar{R}^{c_2}$.

<u>Proof</u>: First of all, to simplify the proof, choose a real matrix K for constructing the matrix a (see (2.1.4)); thus, $a = K\cdot d^T$ is real. Then we have: For $x \in V_\omega^{c_2}$,

$$a\cdot x = 0 \Leftrightarrow a\cdot\mathfrak{J}x = 0. \tag{3.4.11}$$

Indeed, if $x = \alpha \cos \omega t + \beta \sin \omega t$, then $a\cdot x = 0 \Rightarrow a\cdot\alpha = 0$ and $a\cdot\beta = 0$, so that $a\cdot\mathfrak{J}x = a\cdot(\alpha - i\beta) = 0$. The converse is obvious.

Now, suppose that $j \in V_\omega^{c_2}$ is a solution of $\hat{\mathfrak{n}}$ corresponding to $e \in V_\omega^{c_2}$.

By (3.4.11), $\mathfrak{J}j$ satisfies K_2 for h^+.

On the other hand, let $\xi \in \bar{R}^{c_2}$ be such that $a \cdot \xi = 0$. If $\xi = \xi_1 + i\xi_2$, $\xi_1, \xi_2 \in R^{c_2}$, then $a \cdot \xi_1 = 0$, $a \cdot \xi_2 = 0$; consequently, the vectors $\xi_1 \cos \omega t$, $\xi_1 \sin \omega t$, $\xi_2 \cos \omega t$, $\xi_2 \sin \omega t$ are in $V_\omega^{c_2}$ and satisfy the equation $a \cdot x = 0$. However, by K_1 for $\hat{h}$, $\langle c, u_1 \cos \omega t + u_2 \sin \omega t \rangle_{c_2} = 0$ for each $c \in V_\omega^{c_2}$ with $a \cdot c = 0$, where we denoted $\hat{Z}j - e = u_1 \cos \omega t + u_2 \sin \omega t$. Then, taking for c the above four vectors, we get by (3.4.3), $\xi_1^T \cdot u_1 = 0$, $\xi_1^T \cdot u_2 = 0$, $\xi_2^T \cdot u_1 = 0$, $\xi_2^T \cdot u_2 = 0$. Hence, $(\xi_1 + i\xi_2)^T \cdot (u_1 - iu_2) = 0$, i.e., $\langle \xi, u_1 - iu_2 \rangle = 0$. However, by the above assertion, $u_1 - iu_2 = \mathfrak{J}(\hat{Z}j - e) = Z^+(\mathfrak{J}j) - \mathfrak{J}e$, and consequently, $\langle \xi, Z^+(\mathfrak{J}j) - \mathfrak{J}e \rangle = 0$. Thus, $\mathfrak{J}j$ satisfies K_1 for h^+, i.e., $\mathfrak{J}j$ is a solution of h^+ corresponding to $\mathfrak{J}e$.

Conversely, assume that, for a given $e \in V_\omega^{c_2}$, the element $\mathfrak{J}j$ for some $j \in V_\omega^{c_2}$ is a solution of h^+ corresponding to $\mathfrak{J}e$. Since $a \cdot (\mathfrak{J}j) = 0$ by K_2 for h^+, it follows by (3.4.11) that $a \cdot j = 0$, i.e., j satisfies K_2 for $\hat{h}$. Next, choose some $c \in V_\omega^{c_2}$ with $a \cdot c = 0$; then by (3.4.11), $a \cdot (\mathfrak{J}c) = 0$. Thus, by K_1 for h^+, $\langle \mathfrak{J}c, Z^+(\mathfrak{J}j) - \mathfrak{J}e \rangle = 0$. By our above assertion this means that $\langle \mathfrak{J}c, \mathfrak{J}(\hat{Z}j - e) \rangle = 0$, i.e., by (3.4.5), $\langle c, \hat{Z}j - e \rangle_{c_2} = 0$. Hence, j satisfies K_1 for $\hat{h}$, and consequently, j is a solution of $\hat{h}$ corresponding to e.

From Proposition 1 we get immediately:

<u>Proposition 2</u>.

(i) $Q(V_\omega^{c_2}) = V_\omega^{c_2}$ for $\hat{h} \Leftrightarrow Q(\bar{R}^{c_2}) = \bar{R}^{c_2}$ for h^+,

(ii) $\hat{h}$ is regular on $V_\omega^{c_2} \Leftrightarrow h^+$ is regular on $\bar{R}^{c_2}$.

Finally, applying Theorem 2.2.4 and 2.2.2, we get the following result:

<u>Proposition 3</u>. Let the operator $W^+ : \bar{R}^{c_o} \to \bar{R}^{c_o}$ be defined by

$$W^+\alpha = \bar{X}^T \cdot (M_\omega \cdot (X \cdot \alpha)); \tag{3.4.12}$$

then we have for $\hat{h}$,

(a) $Q(V_\omega^{c_2}) = V_\omega^{c_2} \Leftrightarrow W^+\bar{R}^{c_o} = \bar{R}^{c_o}$,

(b) $\hat{h}$ is regular on $V_\omega^{c_2} \Leftrightarrow W^+$ is 1-to-1 on $\bar{R}^{c_o}$.

In this case, the solution $i \in V_\omega^{c_2}$ corresponding to $e \in Q(V_\omega^{c_2})$ is given by

$$i = \mathfrak{J}^{-1}\{X \cdot [W^{+-1}(\bar{X}^T \cdot (\mathfrak{J}e))]\}, \tag{3.4.13}$$

where $W^{+-1} : W^+\bar{R}^{c_o} \to \bar{R}^{c_o}$ is the inverse of W^+.

If, in addition, $\hat{h}$ is finite, then

(c) $Q(V_\omega^{c_2}) = \bar{R}^{c_2}$ and $\hat{n}$ is regular on $V_\omega^{c_2} \Leftrightarrow \det \bar{X}^T \cdot M_\omega \cdot X \neq 0$.

In this case, the solution $i \in V_\omega^{c_2}$ corresponding to $e \in V_\omega^{c_2}$ is given by

$$i = \mathfrak{J}^{-1}\{X\cdot(\bar{X}^T\cdot M_\omega\cdot X)^{-1}\cdot\bar{X}^T\cdot(\mathfrak{J}e)\}. \tag{3.4.14}$$

The proof is obvious.

3.5 A DC-Current Network With One Nonlinear Resistor

Consider the DC-current network $\hat{n}$ sketched in Figure 3.5.

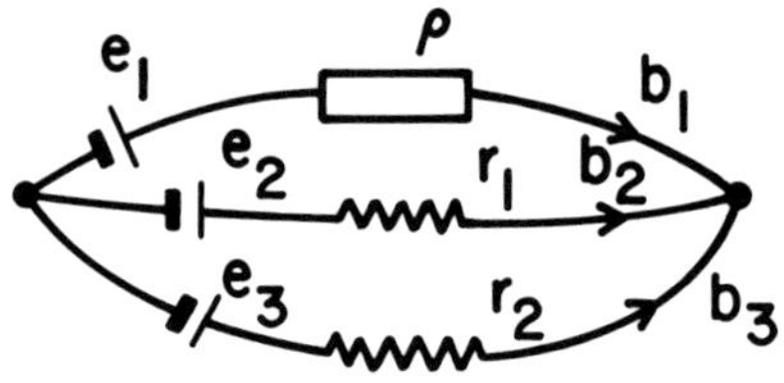

Figure 3.5

Here, the graph G consists of three branches b_1,b_2,b_3. We assume that the branch b_1 contains a resistor described by a function $\rho : R^1 \to R^1$, and the branches b_2,b_3 contain linear resistors r_1 and r_2, respectively. Thus, in this case, $H = R^1$ and

$$\hat{Z} = \operatorname{diag}(\rho(\cdot), r_1\cdot, r_2\cdot). \tag{3.5.1}$$

We are going to show that, if $r_1 + r_2 \neq 0$, the following propositions are true:

Let $\varphi : R^1 \to R^1$ be defined by

$$\varphi(x) = \rho(x) + \frac{r_1 r_2}{r_1+r_2}\, x; \tag{3.5.2}$$

then

(i) $\hat{n}$ is regular on R^3, if and only if φ is 1-to-1 on R^1,

(ii) $Q(R^3) = R^3$, if and only if φ maps R^1 onto R^1.

Indeed, referring to Theorem 2.2.3, $\hat{n}$ is regular on R^3, if and only if the operator $\hat{X}^*\hat{Z}\hat{X}$ is 1-to-1 on R^2 (here, clearly, $c_o = 2$). Similarly, by Theorem 2.2.4, $Q(R^3) = R^3 \Leftrightarrow \hat{X}^*\hat{Z}\hat{X}R^2 = R^2$.

However, since our network is finite, we can easily see that, in the above relations, the operator $\hat{X}$ can be replaced by an operator $\hat{Y}$ generated by a matrix Y whose columns constitute a (not necessarily orthonormal) basis in $N_{\hat{a}}$. As a matter of fact, $X = Y \cdot M$ with M nonsingular, which proves our assertion. (See also Problem 2.2.1.)

Thus, from Figure 3.5, it is apparent that we can put

$$Y = \begin{bmatrix} 1 & 0 \\ -1 & 1 \\ 0 & -1 \end{bmatrix} . \tag{3.5.3}$$

Now, if $x = [x_1, x_2]^T \in R^2$, we get $Y \cdot x = \begin{bmatrix} x_1 \\ -x_1+x_2 \\ -x_2 \end{bmatrix}$, and consequently, by (3.5.1),

$$\hat{Z}\hat{Y}x = \begin{bmatrix} \rho(x_1) \\ -r_1x_1 + r_1x_2 \\ -r_2x_2 \end{bmatrix} ;$$

hence,

$$\hat{Y}*\hat{Z}\hat{Y}x = \begin{bmatrix} \rho(x_1) + r_1x_1 - r_1x_2 \\ -r_1x_1 + r_1x_2 + r_2x_2 \end{bmatrix} . \tag{3.5.4}$$

Denoting the right-hand side of (3.5.4) by $[z_1, z_2]^T$, we get equations

$$\rho(x_1) + r_1x_1 - r_1x_2 = z_1$$

$$-r_1x_1 + r_1x_2 + r_2x_2 = z_2;$$

this system is equivalent to

$$\varphi(x_1) = z_1 + \frac{r_1}{r_1+r_2} z_2, \tag{3.5.5}$$

$$x_2 = \frac{r_1}{r_1+r_2} x_1 + \frac{z_2}{r_1+r_2} .$$

However, from the above equivalences and from (3.5.5) it follows immediately that our propositions (i) and (ii) are true.

Note that the basic idea of this simple example can be generalized. In Section 4.2, using the concept of a "driving-point-set impedance" we will prove a theorem on the existence and uniqueness of a solution of a linear Hilbert network which contains finitely many nonlinear elements.

Problems

3.5.1 Referring to Example 3.2, we can easily verify that the requirement

$i \in D$ (which is enforced by the presence of derivatives in the problem) imposes severe limitations on the existence of a solution. To see this, consider the R, L, C-network with constant elements whose graph G is sketched in the following figure.

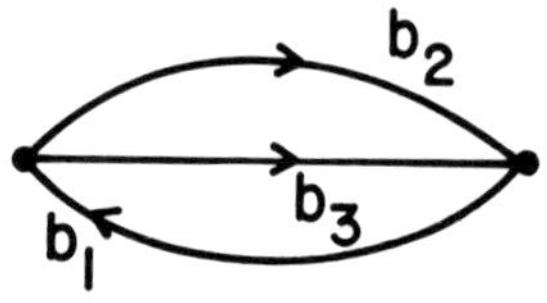

Using the notation of Example 3.2, let L = diag(0,0,2), R = diag(1,2,2) and S = diag(0,1,0), i.e., let the operator $\hat{Z}$ be defined on D by $(\hat{Z}x)(t) = Lx'(t) + Rx(t) + S\int_0^t x(t)dt$; also, assume that $i_o = 0$, i.e., our network is in an equilibrium at t = 0. Now, show that if $e = [1,0,0]^T \in L_2^3[o,T]$, then the network in question does not possess any solution.

Roughly speaking, the trouble is that the requirement that i be absolutely continuous and $i(0) = i_o$ is too restrictive in the given setting. However, we can easily surmount this difficulty by defining "a generalized solution". To explain this idea, let us modify (and generalize, too), the situation considered in Example 3.2 as follows:

Let G be a locally finite oriented graph, and let c_2, a, etc. have the usual meaning. Given $i_o \in R^{c_2}$ with $a \cdot i = 0$, let D be defined as before, i.e., $D = \{x : x = [x_k(t)]$ is a c_2-vector in $L_2^{c_2}[o,T]$, x_k absolutely continuous on $[o,T]$, $x' \in L_2^{c_2}[o,T]$, $x(0) = i_o\}$. Furthermore, assume that

(i) $L(\xi,t) : R^{c_2} \times [o,T] \to R^{c_2}$ is such that, for each $x \in D$, $L(x(t),t)$ is absolutely continuous on $[o,T]$ and $\{L(x(t),t)\}' \in L_2^{c_2}[o,T]$,

(ii) $R(\xi,t) : R^{c_2} \times [o,T] \to R^{c_2}$ is such that $R(x(t),t) \in L_2^{c_2}[o,T]$ for every $x \in L_2^{c_2}[o,T]$.

(iii) $S(\xi,t) : R^{c_2} \times [o,T] \to R^{c_2}$ is such that $S(\int_0^t x(T)dT,t) \in L_2^{c_2}[o,T]$ for every $x \in L_2^{c_2}[o,T]$.

Let $\hat{Z} : D \to L_2^{c_2}[o,T]$ be defined by

$$(\hat{Z}x)(t) = \{L(x(t),t)\}' + R(x(t),t) + S(\int_0^t x(\omega)d\omega,t).$$

A solution $i \in D$ of $\hat{n} = (\hat{Z},G)$ corresponding to $e \in L_2^{c_2}[o,T]$ will be called an elementary solution.

Next, let $\hat{\hat{Z}} : L_2^{c_2}[o,T] \to L_2^{c_2}[o,T]$ be defined by

$$(\hat{\hat{Z}}x)(t) = L(x(t),t) - L(i_o,o) + \int_0^t R(x(\tau),\tau)d\tau$$

$$+ \int_0^t S(\int_0^\tau x(\sigma)d\sigma,\tau)d\tau.$$

Now, we define:

Let $e \in L_2^{c_2}[o,T]$; an element $i \in L_2^{c_2}[o,T]$ will be called a generalized solution of $\hat{n}$ corresponding to e, if i is a solution of $\hat{\hat{n}} = (\hat{\hat{Z}},G)$ corresponding to $\int_0^t e(\tau)d\tau$.

Prove that the following proposition is true:

If $i \in D$ is an elementary solution of $\hat{n}$ corresponding to $e \in L_2^{c_2}[o,T]$, then i is also a generalized solution of $\hat{n}$ corresponding to e.

Observe that a generalized solution is not required to be absolutely continuous and to satisfy the condition $i(0) = i_o$; consequently, the class of generalized solutions is substantially larger than the class of elementary solutions.

Also, show that the specific R, L, C-network discussed above possesses a (unique) generalized solution for any $e \in L_2^3[o,T]$.

CHAPTER 4

DRIVING POINT SET IMPEDANCE OF A HILBERT NETWORK

In this Chapter we will introduce the concept of a driving point set impedance which is a generalization of the driving point impedance known from the classical network theory [6] and prove two theorems giving conditions under which the driving point set impedance is an operator. Then we will consider linear Hilbert networks that contain finitely many nonlinear elements and show that the behavior of such a network is completely determined by the driving point set impedance of branches containing nonlinear elements and by the set mapping describing the latter.

Moreover, we will generalize the Shannon-Hagelbarger theorem [10, 11] to the case of a Hilbert network, i.e, we will show that, in a certain sense, the driving point set impedance $R(\hat{Z})$ of a network $(\hat{Z},G)$ is a concave nondecreasing function of $\hat{Z}$. **Finally**, we will establish the driving point set impedance of some elementary structures and thus generalize the classical rules on series and parallel connections.

4.1 Definition, Conditions Under Which the Driving Point Set Impedance Is An Operator

In the subsequent considerations, we will use the notation employed in the proof of Theorem 2.5.4, i.e. Let $c \leq \aleph_0$ and let n be an integer with $1 \leq n \leq c$. If $x = [x_k] \in H^c$, we put $(x : = [x_1, x_2, \ldots, x_n]^T \in H^n$; if $M \subset H^c$, we put $(N)_n = \{(x)_n : x \in M\}$. Analogously, for $y = [y_1, y_2, \ldots, y_n]^T \in H^n$, we let $(y)' = [y_1, y_2, \ldots, y_n, 0, 0, \ldots]^T \in H^c$; for $N \subset H^n$, we let $(N)' = \{y' : y \in N\}$

Let $\hat{n}$ be a Hilbert network with $F \neq \emptyset$, let A be the admittance of $\hat{n}$ and let $n \leq c_2$ be a positive integer; define the sets

$$Q_n = \{e : e \in H^n, e' \in Q(D)\}, \tag{4.4.1}$$

$$D_n = ((A(Q_n)'))^o)_n \tag{4.1.2}$$

Obviously, D_n is the set of all n-vectors of currents existing in branches $b_1, b_2, \ldots, b_n$ provided the EMF excitations are present only in $b_1, b_2, \ldots, b_n$.

It is clear that if $Q_n \neq \emptyset$, then $D_n \neq \emptyset$; also note that $Q_n \subset H^n$ and $D_n \subset H^n$.

Definition 1. Let $\hat{n}$ be a Hilbert network with $F \neq \emptyset$, $D_n \neq \emptyset$, and let $A : Q(D) \to \mathfrak{S}(N_a \hat{} \cap D)$ be the admittance of $\hat{n}$. Then the quasiinverse R of the set mapping $A_n : Q_n \to \mathfrak{S}(D_n)$, defined by

$$A_n e = (A(e)')_n, \qquad (4.1.3)$$

will be called the driving point set impedance of branches $b_1, b_2, \ldots, b_n$ (further DPSI). If $n = 1$, the DPSI will also be called the driving point impedance.

A comment on this definition is in order. First note that, in the present context, the DPSI of $b_1, b_2, \ldots, b_n$ exists for any Hilbert network satisfying the requirements $F \neq \emptyset$ and $D_n \neq \emptyset$. Also observe that we do not lose any generality by confining our attention to branches $b_1, b_2, \ldots, b_n$ only, since the enumeration of the branches in G is immaterial.

The physical meaning of the DPSI is clarified by the following fact:

Proposition. Let $j \in D_n$; then for any $e \in Rj$ there exists a solution i of $\hat{n}$ corresponding to $(e)'$ such that $(i)_n = j$.

Proof: Choose some $e \in Rj$. Since $R : D_n \to \mathfrak{S}(Q_n)$, we have $e \in Q_n$, and consequently, by Lemma 1.1.1, $j \in R^- e = (A_n^-)^- e = A_n e = (A(e)')_n$. This means that there exists $i \in A(e)'$ such that $j = (i)_n$. However, since $A(e)'$ is the set of all solutions corresponding to $(e)'$ by Theorem 2.2.2, our proof is complete.

The most important situation occurs when both the admittance and DPSI are operators. In this case, the above proposition reads: "If $j \in D_n$, then the unique solution i of $\hat{n}$ corresponding to $(Rj)'$ has the property that $(i)_n = j$". In order to investigate this case more closely, we introduce several new concepts and carry out some auxiliary considerations.

Let G be a locally finite oriented graph; a nonzero vector $\xi = [\xi_k] \in R^{c_2}$ will be called a loop, if $a \cdot \xi = 0$ and each element ξ_k attains one of values -1, 0, 1.

Note that every loop is a simple vector, i.e., all but finitely many of its entries are zero.

We will say that a loop $\xi = [\xi_k]$ contains (does not contain) a branch b_j, if $\xi_j \neq 0$ ($\xi_j = 0$).

Remark 1. In the above definition we deviate from the standard terminology of graph theory. This is done only for the sake of simplicity of presentation and can hardly lead to misunderstanding.

Definition 2. Let G be a locally finite oriented graph, and let n be an integer with $1 \leq n \leq c_o$. The set of branches $\{b_1,b_2,\ldots,b_n\}$ will be called regular, if there exist loops $\xi^1,\xi^2,\ldots,\xi^n$ such that, for each $k = 1,2,\ldots,n$, the loop ξ^k contains b_k and does not contain any other branch in the set $\{b_1,b_2,\ldots,b_n\}$.

From this definition it follows immediately that vectors $\xi^1,\xi^2,\ldots,\xi^n$ are linearly independent.

Lemma 4.1.1. Let the set of branches $\{b_1,b_2,\ldots,b_n\}$ be regular, let X be any $c_2 \times c_o$-matrix whose columns constitute an orthonormal basis in the solution space of the equation $a\cdot\xi = 0$, $\xi \in R^{c_2}$, and let H be a separable Hilbert space. Then there exists $\lambda > 0$ such that for the operator $\hat{X} : H^{c_o} \to H^{c_2}$ defined by $\hat{X}z = X\cdot z$ we have

$$\|\hat{X}^*(z)'\|_{c_o} \geq \lambda\|z\|_n \qquad (4.1.4)$$

for every $z \in H^n$.

Proof: Let $\{\xi^1,\xi^2,\ldots,\xi^n\}$ be the linearly independent set of loops existing by the above definition of regularity, and let $\{\xi_o^1,\xi_o^2,\ldots,\xi_o^n\}$ be the orthonormal set in R^{c_2} obtained from the former by applying the Gram-Schmidt process. From the properties of ξ^k's it follows that the $n \times n$-matrix M_{11} formed by the n first rows of $[\xi_o^1 \vdots \xi_o^2 \vdots \ldots \vdots \xi_o^n]$, is an upper triangular matrix such that each element in the main diagonal is nonzero; consequently, M_{11} is nonsingular. Thus, it is clear that there exists $\mu > 0$ such that

$$\|\bar{M}_{11}^T\cdot\nu\|_n \geq \mu\|\nu\|_n \qquad (4.1.5)$$

for every $\nu \in H^n$.

Next, choose vectors $\xi_o^k \in R^{c_2}$, $k = n+1, n+2,\ldots$ so that $\{\xi_o^1,\xi_o^2,\ldots,\xi_o^n,\xi_o^{n+1},\ldots\}$ constitutes an orthonormal basis in the solution space of $a\cdot\xi = 0$, $\xi \in R^{c_o}$, and define the $c_2 \times c_o$ matrix X_o by $X_o = [\xi_o^1 \vdots \xi_o^2 \vdots \ldots]$. Thus, decomposing X_o into blocks, we can write

$$X_o = \left[\begin{array}{c|c} M_{11} & M_{12} \\ \hline M_{21} & M_{22} \end{array}\right] \quad . \qquad (4.1.6)$$

Now, choosing $z \in H^n$ we get by Lemma 2.1.2 (v), (4.1.6) and (4.1.5),

$$\|\hat{X}_o^*(z)'\|^2_{c_o} = \|\bar{X}_o^T\cdot(z)'\|^2_{c_o} = \|\bar{M}_{11}^T\cdot z\|^2_n + \|\bar{M}_{12}^T\cdot z\|^2_{c_o-n} \qquad (4.1.7)$$
$$\geq \|\bar{M}_{11}^T\cdot z\|^2_n \geq \mu^2\|z\|^2_n.$$

To conclude the proof, choose a $c_2 \times c_o$ matrix X whose columns constitute an orthonormal basis in the solution space of $a\cdot\xi = 0$, $\xi \in R^{c_2}$, and let $\hat{X} : H^{c_o} \to H^{c_2}$ be the operator generated by X. Define the operator $S : H^{c_o} \to H^{c_o}$ by

$$S = \hat{X}_o^* \hat{X}. \qquad (4.1.8)$$

By Lemma 2.1.2 (i), S is a linear bounded operator. Moreover, S maps H^{c_o} onto H^{c_o}. Indeed, by Lemma 2.2.2 (i), $\hat{X}H^{c_o} = N_{\hat{a}}$, and consequently, $SH^{c_o} = \hat{X}_o^*\hat{X}H^{c_o} = \hat{X}_o^*N_{\hat{a}}$; however, because also $\hat{X}_oH^{c_o} = N_{\hat{a}}$, it follows by Lemma 2.1.2 (v) that $H^{c_o} = \hat{X}_o^*\hat{X}_oH^{c_o} = \hat{X}_o^*N_{\hat{a}}$. Hence, $SH^{c_o} = H^{c_o}$.

Furthermore, it is easy to see that, on H^{c_o},

$$\hat{X}_oS = \hat{X}. \qquad (4.1.9)$$

Indeed, choose an $x \in H^{c_o}$. Then $\hat{X}_oSx = \hat{X}_o\hat{X}_o^*\hat{X}x$; however, since $\hat{X}x \in N_{\hat{a}}$ by Lemma 2.1.2 (i) and $\hat{X}_o\hat{X}_o^*$ is the orthogonal projection onto $N_{\hat{a}}$ by Lemma 2.1.2 (vii), we have $\hat{X}_o\hat{X}_o^*\hat{X}x = \hat{X}x$, which confirms (4.1.9).

Finally, it follows that S is 1-to-1. Indeed, assume that $Sx = 0$ for some $x \in H^{c_o}$; then, by (4.1.9), $\hat{X}_oSx = \hat{X}x = 0$, and consequently, by Lemma 2.1.2 (i), $x = 0$.

Summarizing these facts we see that S is a bounded bijection; thus, by the open mapping theorem, S^{-1} is bounded. Consequently, $(S^*)^{-1} = (S^{-1})^*$ is bounded, i.e., there exists a $\mu' > 0$ such that

$$\|S^*x\|_{c_o} \geq \mu'\|x\|_{c_o} \qquad (4.1.10)$$

for each $x \in H^{c_o}$.

To finish the proof, let $z \in H^n$; then (4.1.9), (4.1.10) and (4.1.7) yield $\|\hat{X}^*(z)'\|_{c_o} = \|S^*\hat{X}_o^*(z)'\|_{c_o} \geq \mu'\|\hat{X}_o^*(z)'\|_{c_o} \geq \mu'\mu\|z\|_n$ which is the inequality (4.1.4)

Now we are ready to state conditions under which the DPSI of a network is an operator.

Theorem 4.1.1. Let $\hat{\mathfrak{n}} = (\hat{Z},G)$ be a Hilbert network, let $\hat{Z}$ be an operator on $D \subset H^{c_2}$, let $F \neq \emptyset$, $Q_n \neq \emptyset$ and let the set of branches $\{b_1,b_2,\ldots,b_n\}$ be regular. Furthermore, assume that

(i) there exist $c > 0$ and $p > 1$ such that

$$\mathrm{Re}\langle Wx_1-Wx_2,x_1-x_2\rangle_{c_o} \geq c\|x_1-x_2\|_{c_o}^{p} \qquad (4.1.11)$$

for all $x_1,x_2 \in F$, where $W = \hat{X}^*\hat{Z}\hat{X} : F \to WF$,

(ii) there exist $d > 0$ and $0 < q \leq 1$ such that

$$\|Wx_1-Wx_2\|_{c_o} \leq d\|x_1-x_2\|_{c_o}^{q} \qquad (4.1.12)$$

for all $x_1,x_2 \in F$.

Then the DPSI $R : D_n \to Q_n$ of branches $b_1,b_2,\ldots,b_n$ is an operator, and R has the properties:

$$\|Rj_1-Rj_2\|_n \leq \gamma\|j_1-j_2\|_n^{\frac{q}{p-q}}, \quad \gamma > 0 \qquad (4.1.13)$$

for all $j_1,j_2 \in D_n$,

$$\mathrm{Re}\langle Rj_1-Rj_2,j_1-j_2\rangle_n \geq c\|j_1-j_2\|_n^{p} \qquad (4.1.14)$$

for all $j_1,j_2 \in D_n$.

Proof: First, Theorem 2.5.2 implies that $\hat{\mathfrak{n}}$ is regular on D and its admittance (operator) $A : Q(D) \to D$ satisfies the inequality

$$\mathrm{Re}\langle Ax_1-Ax_2,x_1-x_2\rangle_{c_2} \geq \alpha\|\hat{X}^*(x_1-x_2)\|_{c_o}^{\lambda} \qquad (4.1.15)$$

for all $x_1,x_2 \in Q(D)$, where $\lambda = \frac{p}{q}$ and $\alpha = cd^{-\lambda} > 0$. Observe that, by our hypothesis, $\lambda > 1$.

Let the operator $A_n : Q_n \to D_n$ be defined by (4.1.3). From (4.1.1) and (4.1.2) it follows that A_n maps Q_n onto D_n. We are going to show that A_n is 1-to-1 on Q_n. Indeed, if $e_1,e_2 \in Q_n$, we have by (4.1.3) and (4.1.15),

$$\mathrm{Re}\langle A_ne_1-A_ne_2,e_1-e_2\rangle_n = \mathrm{Re}\langle A(e_1)'-A(e_2)',(e_1-e_2)'\rangle_{c_2} \qquad (4.1.16)$$
$$\geq \alpha\|\hat{X}^*(e_1-e_2)'\|_{c_o}^{\lambda}.$$

However, invoking Lemma 4.1.1 we get from (4.1.16)

$$\mathrm{Re}\langle A_n e_1 - A_n e_2, e_1 - e_2\rangle \geq \alpha\mu^\lambda \|e_1 - e_2\|_n^\lambda \tag{4.1.17}$$

with $\mu > 0$. This inequality shows readily that A_n is 1-to-1. Consequently, the quasiinverse of A_n coincides with A_n^{-1}, i.e., the DPSI $R = A_n^{-1} : D_n \to Q_n$ is an operator.

To prove the inequality (4.1.13) we can proceed as follows: The Schwartz inequality and (4.1.17) imply that for $e_1, e_2 \in Q_n$,

$$\|A_n e_1 - A_n e_2\|_n \geq \alpha\mu^\lambda \|e_1 - e_2\|_n^{\lambda-1}. \tag{4.1.18}$$

Choose $j_1, j_2 \in D_n$ and put $e_k = Rj_k = A_n^{-1} j_k$, $k = 1,2$. Then (4.1.18) yields $\|j_1 - j_2\|_n \geq \alpha\mu^\lambda \|Rj_1 - Rj_2\|_n^{\lambda-1}$, i.e.,

$$\|Rj_1 - Rj_2\|_n \leq \gamma \|j_1 - j_2\|_n^{\frac{1}{\lambda-1}}, \quad \gamma > 0.$$

This, however, is (4.1.13), since $\frac{1}{\lambda-1} = \frac{q}{p-q}$.

Finally, to prove (4.1.14), choose $j_1, j_2 \in D_n$, and put $e_k = (Rj_k)' \in Q(D)$, $k = 1,2$; also, let i_1 and i_2 be the solution of $\hat{n}$ corresponding to e_1 and e_2, respectively. Then we have

$$\mathrm{Re}\langle Rj_1 - Rj_2, j_1 - j_2\rangle_n = \mathrm{Re}\langle e_1 - e_2, i_1 - i_2\rangle_{c_2}. \tag{4.1.19}$$

On the other hand, by the definition of a solution we have (note that $\hat{Z}$ is an operator),

$$\langle e_1 - e_2, i_1 - i_2\rangle_{c_2} = \langle e_1 - e_2, Ae_1 - Ae_2\rangle_{c_2} = \langle \hat{Z}i_1 - \hat{Z}i_2, i_1 - i_2\rangle_{c_2}.$$

Hence, by (4.1.19)

$$\mathrm{Re}\langle Rj_1 - Rj_2, j_1 - j_2\rangle_n = \mathrm{Re}\langle \hat{Z}i_1 - \hat{Z}i_2, i_1 - i_2\rangle_{c_2}. \tag{4.1.20}$$

However, (4.1.11) is equivalent to the condition

$$\mathrm{Re}\langle \hat{Z}y_1 - \hat{Z}y_2, y_1 - y_2\rangle_{c_2} \geq c\|y_1 - y_2\|_{c_2}^p \tag{4.1.21}$$

for all $y_1, y_2 \in N_{\hat{a}} \cap D$. Since $i_1, i_2 \in N_{\hat{a}} \cap D$, we get from (4.1.20),

$$\mathrm{Re}\langle Rj_1 - Rj_2, j_1 - j_2\rangle_n \geq c\|i_1 - i_2\|_{c_2}^p. \tag{4.1.22}$$

Since $(i_1)_n = j_1$, $(i_2)_n = j_2$, it follows that $\|i_1 - i_2\|_{c_2} \geq \|j_1 - j_2\|_n$; this together with (4.1.22) concludes the proof of (4.1.14).

From Theorem 4.1.1 we get readily the following result:

Theorem 4.1.2. Let H be a real Hilbert space, let $\hat{\mathfrak{N}} = (\hat{Z},G)$ be a Hilbert network with $\hat{Z}$ being an operator on H^{c_2}, and let the set of branches $\{b_1,b_2,\ldots,b_n\}$ be regular. Assume that

(i) there exist $c > 0$ and $p > 1$ such that

$$\langle Wx_1-Wx_2,x_1-x_2\rangle_{c_o} \geq c\|x_1-x_2\|_{c_o}^p \tag{4.1.23}$$

for all $x_1,x_2 \in H^{c_o}$, where $W = \hat{X}^*\hat{Z}\hat{X} : H^{c_o} \to H^{c_o}$,

(ii) there exist $d > 0$ and $0 < q \leq 1$ such that

$$\|Wx_1-Wx_2\|_{c_o} \leq d\|x_1-x_2\|_{c_o}^q \tag{4.1.24}$$

for all $x_1,x_2 \in H^{c_o}$.

Then the DPSI R of branches $b_1,b_2,\ldots,b_n$ is an operator from H^n onto H^n and (4.1.13), (4.1.14) holds for all $j_1,j_2 \in H^n$.

Proof: Since W is continuous by (ii), Theorem 2.3.2 implies that $Q(H^{c_2}) = H^{c_2}$. Consequently, by (4.1.1), $Q_n = H^n$. Thus, by (4.1.17), the operator $A_n : H^n \to D_n \subset H^n$ satisfies the condition

$$\langle A_ne_1-A_ne_2,e_1-e_2\rangle_n \geq \beta\|e_1-e_2\|_n^\lambda \tag{4.1.25}$$

for all $e_1,e_2 \in H^n$, where $\beta > 0$ and $\lambda > 1$.

On the other hand, if $e_1,e_2 \in H^n$, we have by the definition (4.1.3) and Theorem 2.2.5 (c), (we use the inequality (2.2.15))

$$\begin{aligned} \|A_ne_1-A_ne_2\|_n &\leq \|A(e_1)'-A(e_2)'\|_{c_2} \\ &\leq c^{-\frac{1}{p-1}}\|\hat{X}^*(e_1-e_2)'\|_{c_o}^{\frac{1}{p-1}} \leq c^{-\frac{1}{p-1}}\|(e_1-e_2)'\|_{c_2}^{\frac{1}{p-1}} \\ &= c^{-\frac{1}{p-1}}\|e_1-e_2\|_n^{\frac{1}{p-1}}. \end{aligned} \tag{4.1.26}$$

However, (4.1.26) shows that A_n is continuous on H^n. Moreover, (4.1.25) implies (see the proof of Theorem 1.2.7) that A_n is coercive and monotone on H^n. Hence, by continuity, A_n is maximal monotone on H^n, and consequently, $A_nH^n = H^n$ by Lemma 1.1.6, i.e., $D_n = H^n$. The rest of the proof follows from Theorem 4.1.1.

4.2 Linear Hilbert Networks Containing Finitely Many Nonlinear Elements

In this Section we will discuss the behavior of linear Hilbert networks that contain finitely many nonlinear elements.

Theorem 4.2.1. Let $\hat{n}' = (\hat{Z}',G)$ be a Hilbert network with $\hat{Z}'$ being a linear (not necessarily bounded) operator defined on a linear subspace $D' \subset H^{c_2}$. Assume that

(i) $\hat{n}'$ is regular on D' and $Q(D') = H^{c_2}$,

(ii) the DPSI of branches $b_1, b_2, \ldots, b_n$ is an operator $R : D_n \to H^n$, where $D_n \subset H^n$ is defined by (4.1.2),

(iii) for any solution i' of $\hat{n}'$ we have $(i')_n \in D_n$.

Furthermore, let $Z^+ : D^+ \to \mathfrak{S}(H^n)$ be a set mapping, where $D^+ \subset H^n$, and let the set mapping $\hat{Z}''$ be defined on $D'' = \{x : x \in H^{c_2}, (x)_n \in D^+\}$ by

$$\hat{Z}''x = (Z^+(x)_n)'. \qquad (4.2.1)$$

Let $\hat{n} = (\hat{Z},G)$, where $\hat{Z} = \hat{Z}' + \hat{Z}''$. Then

(a) $\hat{n}$ is regular on $D' \cap D'' \Leftrightarrow R + Z^+$ is simple on $D_n \cap D^+$,

(b) for $\hat{n}$, $Q(D' \cap D'') = H^{c_2} \Leftrightarrow [(R+Z^+)(D_n \cap D^+)]^o = H^n$.

A comment on the physical meaning of this theorem is in order. It is clear that, by definition of $\hat{Z}$, the network $\hat{n}$ is obtained from the linear network $\hat{n}'$ by inserting additional nonlinear elements into the branches $b_1, b_2, \ldots, b_n$. The behavior of these elements is described by the set mapping Z^+. Since we do not make any special assumption about Z^+, mutual couplings may exist between the additional nonlinear elements (but no couplings are allowed between the nonlinear elements and elements of the original network $\hat{n}'$). In view of propositions (a) and (b), the regularity of the composite network $\hat{n}$ and the existence of a solution of $\hat{n}$ for a given excitation by EMF's is completely determined by the behavior of the set mapping $R + Z^+$, where R is the DPSI of those branches in $\hat{n}'$, which contain the nonlinear elements.

Proof of Theorem 4.2.1. First of all, from the assumption $Q(D') = H^{c_2}$ and (4.1.1) it follows that $Q_n = H^n$. Moreover, since $\hat{Z}'$ is a linear operator and $\hat{n}'$ is regular on D', it follows that the admittance $A' : H^{c_2} \to D'$ of $\hat{n}'$ is a linear operator. (Witness Theorems 2.2.2 and 2.2.3; $W = \hat{X}^*\hat{Z}'\hat{X}$ is a linear operator on H^{c_o} and so is $W^{-1} = W^-$.) Thus, by (4.1.2), D_n is a linear subspace of H^n.

Since $n < c_2$ we will introduce the following notation: if $x = [x_k] \in H^{c_2}$, we let $(x)_{-n} = [x_{n+1}, x_{n+2}, \ldots]^T \in H^{c_2-n}$ (here we put $c_2 - n = \aleph_o$ if $c_2 = \aleph_o$). Then we clearly have $x = [(x)_n^T \vdots (x)_{-n}^T]^T$ for any $x \in H^{c_2}$.

Due to linearity of A' it follows that there exist linear operators

$A_{11}: H^n \to H^n$, $A_{12}: H^{c_2-n} \to H^n$, $A_{21}: H^n \to H^{c_2-n}$ and $A_{22}: H^{c_2-n} \to H^{c_2-n}$ such that

$$A'x = \left[\begin{array}{c|c} A_{11} & A_{12} \\ \hline A_{21} & A_{22} \end{array}\right] \cdot \left[\begin{array}{c} (x)_n \\ \hline (x)_{-n} \end{array}\right] \qquad (4.2.2)$$

for every $x \in H^{c_2}$. Moreover, the assumption (iii) implies that we even have $A_{11}: H^n \to D_n$ and $A_{12}: H^{c_2-n} \to D_n$.

On the other hand, by (4.1.3) and (4.2.2) we have for any $z \in Q_n = H^n$,

$$A_n z = (A(z)')_n = A_{11} z. \qquad (4.2.3)$$

Since, by (4.1.3) and (4.1.2) $A_n H^n = D_n$, and by (ii) the DPSI $R: D_n \to H^n$ is an operator, (4.2.3) shows that $A_n = A_{11}$, i.e., the inverse $A_{11}^{-1}: D_n \to H^n$ exists and $A_{11}^{-1} = R$.

Now, we are going to show that the following equivalence ($\mathcal{E}$) is true:

($\mathcal{E}$) Let $i \in N_{\hat{A}} \cap (D' \cap D'')$ and let $e \in H^{c_2}$; then i is a solution of $\hat{\Pi}$ corresponding to $e \Leftrightarrow$ there exists an element $\tilde{\nu} \in Z^+(i)_n$ such that i is a solution of $\hat{\Pi}'$ corresponding to $e-(\tilde{\nu})'$. In this case,

$$(e)_n + RA_{12}(e)_{-n} \in (R+Z^+)(i)_n, \qquad (4.2.4)$$

$(i)_n \in D_n \cap D^+$, and the element $\tilde{\nu}$ is determined uniquely.

Indeed, assume first that $i \in N_{\hat{A}} \cap (D' \cap D'')$ is a solution of $\hat{\Pi}$ corresponding to $e \in H^{c_2}$; then, since $i \in D''$, we have $(i)_n \in D^+$. Moreover, by condition K_1 in the definition of a solution, there exists $\nu \in Zi$ such that

$$\nu - e \in N_{\hat{A}}^{\perp}. \qquad (4.2.5)$$

However, since $\hat{Z} = \hat{Z}' + \hat{Z}''$, we have by (4.2.1), $\hat{Z}i = \hat{Z}'i + \hat{Z}''i = \hat{Z}'i + (Z^+(i)_n)'$; this amounts to saying that there exists $\tilde{\nu} \in Z^+(i)_n$ such that $\nu = \hat{Z}'i + (\tilde{\nu})'$. Consequently, (4.2.5) can be written as

$$\hat{Z}'i - (e-(\tilde{\nu})') \in N_{\hat{A}}^{\perp}, \qquad (4.2.6)$$

i.e., because $i \in N_{\hat{A}} \cap D'$, i is a solution of $\hat{\Pi}'$ corresponding to $e-(\tilde{\nu})'$.

On the other hand, by (4.2.2) we have

$$(i)_n = A_{11}((e)_n - \tilde{\nu}) + A_{12}(e)_{-n}. \qquad (4.2.7)$$

However, since both elements $A_{11}((e)_n - \tilde{\nu})$ and $A_{12}(e)_{-n}$ are in D_n, (4.2.7) is equivalent to

$$R(i)_n + \tilde{\nu} = (e)_n + RA_{12}(e)_{-n}. \tag{4.2.8}$$

This relation shows that the element $\tilde{\nu}$ is determined uniquely. Moreover, since $\tilde{\nu} \in Z^+(i)_n$, we have $R(i)_n + \tilde{\nu} \in (R+Z^+)(i)_n$; thus, (4.2.8) yields immediately the relation (4.2.4).

Conversely, assume that $i \in N_a\hat{\cap}(D'\cap D'')$ and that there exists $\tilde{\nu} \in Z^+(i)_n$ such that i is a solution of $\hat{\mathfrak{n}}'$ corresponding to $e-(\hat{\tilde{\nu}})'$. Then (4.2.6) holds (it is K_1 for $\hat{\mathfrak{n}}'$), and letting $\nu = \hat{Z}'i + (\tilde{\nu})'$, we see readily that $\nu \in \hat{Z}i$ and (4.2.5) is satisfied. Hence, i is a solution of $\hat{\mathfrak{n}}$ corresponding to e, and our equivalence ($\mathcal{E}$) is proven.

(a) Assume now that $R + Z^+$ is simple on $D_n\cap D^+$. Suppose that, for some $e \in H^{c_2}$, $\hat{\mathfrak{n}}$ has solutions $i, \tilde{i} \in N_a\hat{\cap}(D'\cap D'')$ which correspond to e. Denote $f = (e)_n + RA_{12}(e)_{-n}$; then we have by (4.2.4), $f \in (R+Z^+)(i)_n$ and $f \in (R+Z^+)(\tilde{i})_n$. Consequently, $((R+Z^+)(i)_n) \cap ((R+Z^+)(\tilde{i})_n) \neq \emptyset$, so that, by simplicity of $R+Z^+$, $(i)_n = (\tilde{i})_n$.

Moreover, for i there exists a unique element $\tilde{\nu} \in Z^+(i)_n$, and for $\tilde{i}$ a unique element $\tilde{\tilde{\nu}} \in Z^+(\tilde{i})_n$ such that $\tilde{\nu}$ and $\tilde{\tilde{\nu}}$ have properties described in ($\mathcal{E}$). Since $(i)_n = (\tilde{i})_n$, (4.2.8) shows that necessarily $\tilde{\nu} = \tilde{\tilde{\nu}}$. Now, again by ($\mathcal{E}$), both i and $\tilde{i}$ are solutions of $\hat{\mathfrak{n}}'$ corresponding to the same element $e-(\tilde{\nu})'$. However, since $\hat{\mathfrak{n}}'$ is regular on D' by (i), it is regular on $D'\cap D''$, too; hence, $i = \tilde{i}$, i.e., $\hat{\mathfrak{n}}$ is regular on $D'\cap D''$.

Conversely, assume that $\hat{\mathfrak{n}}$ is regular on $D'\cap D''$. Suppose that there exist $j, \tilde{j} \in D_n\cap D^+$ such that $((R+Z^+)j)\cap((R+Z^+)\tilde{j}) \neq \emptyset$. Then there exists $f \in H^n$ such that $f \in (R+Z^+)j$ and $f \in (R+Z^+)\tilde{j}$. (Note that $R+Z^+$ maps $D_n\cap D^+$ into $\mathfrak{S}(H^n)$). Stated differently, there exists $\tilde{\nu} \in Z^+j$ and $\tilde{\tilde{\nu}} \in Z^+\tilde{j}$ such that

$$f = Rj + \tilde{\nu}, \quad f = R\tilde{j} + \tilde{\tilde{\nu}}. \tag{4.2.9}$$

Let $e = (f-\tilde{\nu})' = (f)'-(\tilde{\nu})'$ and $\tilde{e} = (f-\tilde{\tilde{\nu}})' = (f)'-(\tilde{\tilde{\nu}})'$; since $\tilde{\nu}, \tilde{\tilde{\nu}} \in H^n$, we have $e, \tilde{e} \in H^{c_2}$. Now, since $\hat{\mathfrak{n}}'$ is regular on D' and $Q(D') = H^{c_2}$ by our hypothesis (i), then for e and $\tilde{e}$ there exists a unique solution i and $\tilde{i}$ of $\hat{\mathfrak{n}}'$ corresponding to e and $\tilde{e}$, respectively. On the other hand, by (4.2.2) and (4.2.9) we have $(i)_n = A_{11}(f-\tilde{\nu}) = A_{11}Rj = j$ (clearly, $(e)_{-n} = ((f-\tilde{\nu})')_{-n} = 0$), and similarly, $(\tilde{i})_n = A_{11}(f-\tilde{\tilde{\nu}}) = A_{11}R\tilde{j} = \tilde{j}$. Also, note that since $i, \tilde{i} \in D'$, and the elements $(i)_n = j$ and $(\tilde{i})_n = \tilde{j}$ belong to D^+, we have $i, \tilde{i} \in N_a\hat{\cap}(D'\cap D'')$. Thus, invoking ($\mathcal{E}$), it follows that i is a solution of $\hat{\mathfrak{n}}$ corresponding to $(f)'$, and $\tilde{i}$ is a solution of $\hat{\mathfrak{n}}$ corresponding to $(f)'$. Hence, by our hypothesis, $i = \tilde{i} \Rightarrow j = (i)_n = (\tilde{i})_n = \tilde{j}$. Thus, $R+Z^+$ is simple on $D_n \cap D^+$, which finishes the proof of the assertion (a).

(b) Assume first that $[(R+Z^+)(D_n \cap D^+)]^o = H^n$. Choose arbitrarily $e \in H^{c_2}$ and construct the element $(e)_n + RA_{12}(e)_{-n} \in H^n$. (This is possible, since $A_{12}(e)_{-n} \in D_n$ and $R : D_n \to H^n$.) By our hypothesis, there exists $j \in D_n \cap D^+$ such that $(e)_n + RA_{12}(e)_{-n} \in (R+Z^+)j$. Thus, there exists $\tilde{\nu} \in Z^+j$ such that

$$(e)_n + RA_{12}(e)_{-n} = Rj + \tilde{\nu}. \tag{4.2.10}$$

On the other hand, by the assumption (i) on $\hat{n}'$, there exists a unique solution $i \in D'$ of $\hat{n}'$ corresponding to $e-(\tilde{\nu}) \in H^{c_2}$. Note that $(i)_n \in D_n$. Moreover, by (4.2.2) and (4.2.10) we have

$$\begin{aligned}(i)_n &= A_{11}(e-\tilde{\nu})')_n + A_{12}(e-(\tilde{\nu})')_{-n} \qquad (4.2.11)\\ &= A_{11}(e)_n - A_{11}\tilde{\nu} + A_{12}(e)_{-n} = A_{11}(Rj-RA_{12}(e)_{-n})\\ &+ A_{12}(e)_{-n} = j.\end{aligned}$$

Since $(i)_n = j \in D^+$, we have $i \in N_{\hat{a}} \cap (D' \cap D'')$, (see the definition of D''). Thus, by $(\mathcal{E})$, i is a solution of $\hat{n}$ corresponding to e. Hence, $Q(D' \cap D'') = H^{c_2}$.

Conversely, assume that $Q(D' \cap D'') = H^{c_2}$, and choose arbitrarily $f \in H^n$. Then, by our hypothesis, there exists a solution $i \in N_{\hat{a}} \cap (D' \cap D'')$ of $\hat{n}$ corresponding to $(f)' \in H^{c_2}$. Note that $(i)_n \in D^+$. However, by $(\mathcal{E})$, i is also a solution of $\hat{n}'$ corresponding to $(f)'-(\tilde{\nu})'$, where $\tilde{\nu} \in Z^+(i)_n$. Consequently, by (iii), $(i)_n \in D_n$, so that $(i)_n \in D_n \cap D^+$. Moreover, by (4.2.4), $((f)')_n + RA_{12}((f'))_{-n} = f \in (R+Z^+)(i)_n$, i.e., $f \in [(R+Z^+)(D_n \cap D^+]^o$. Hence, $[(R+Z^+)(D_n \cap D^+)]^o = H^n$ which completes the proof of proposition (b).

<u>Remark 1</u>. (a) It is easy to see that the concept of the DPSI can be extended without any essential difficulties to the case that countably many branches are involved, and that results like Theorem 4.1.1, 4.1.2 and 4.2.1 can be proved. However, we omit the details.

(b) The concept of the DPSI and Theorem 4.2.1 do not make any essential use of the topological properties of the spaces involved, i.e., they have a purely set-theoretical character. Consequently, Theorem 4.2.1 remains true without any change for algebraic networks considered in Part 6.

Let us now discuss some simple applications of Theorem 4.2.1 in the case $n = 1$.

Example 1. Let G be a finite oriented graph having branches $b_1, b_2, \ldots, b_{c_2}$, and let $\hat{\eta}' = (\hat{Z}', G)$ be a (finite) network built up from constant (not necessarily nonnegative) R, L, C elements, i.e., the operator $\hat{Z}'$ is described by a $c_2 \times c_2$ matrix $[Z'_{ik}]$, where

$$(Z_{ik}x)(t) = L_{ik}x'(t) + R_{ik}x(t) + S_{ik}\int_0^t x(\sigma)d\sigma, \qquad (4.2.12)$$

and L_{ik}, R_{ik}, S_{ik} are real numbers.

For the underlying space H let us take the real space $L_2[0,\tau]$, $\tau > 0$; also let $\mathcal{K}$ be the space of all absolutely continuous functions x on $[0,\tau]$ such that $x(0) = 0$ and $x' \in L_2[0,\tau]$. Clearly, $\mathcal{K} \subset L_2[0,\tau]$.

Now, let $D' = M_1 \times M_2 \times \ldots \times M_{c_2}$, where $M_k = L_2[0,\tau]$, if the branch b_k does not contain any inductance, i.e., if $L_{kj} = 0$ for $j = 1,2,\ldots,c_2$, and $M_k = \mathcal{K}$ in the opposite case.

It is clear that then $\hat{Z}'$ is well-defined on D', and maps D' into $L_2^{c_2}[0,\tau]$.

We will assume that

(1) $M_1 = L_2[0,\tau]$,

(2) $\hat{\eta}'$ is regular on D' and $Q(D') = H^{c_2}$.

If the admittance $\hat{A}' : H^{c_2} \to D'$ of $\hat{\eta}'$ is described by a matrix $[A_{ik}]$ (of type $c_2 \times c_2$), then, as known from the elementary network analysis, each operator A_{ik} has necessarily the form (witness assumption (2))

$$(A_{ik}x)(t) = a_{ik}x(t) + \int_0^t K_{ik}(t-\sigma)x(\sigma)d\sigma, \qquad (4.2.13)$$

where a_{ik} is a real number and $K_{ik}(\sigma)$ is a continuous function on $[0,\infty)$ (in fact, K_{ik} is a linear combination of functions $e^{\lambda t}\{P(t)\cos\omega t + Q(t)\sin\omega t\}$, P, Q being polynomials.)

Furthermore, we will assume that

(3) $a_{11} \neq 0$.

Then it is easy to see that A_{11} is 1-to-1 from $L_2[0,\tau]$ onto itself (it is a Volterra operator), and consequently, $R = A_{11}^{-1} : L_2[0,\tau] \to L_2[0,\tau]$ is the driving point impedance of the branch b_1. Also, R has the form

$$(Rx)(t) = rx(t) + \int_0^t K(t-\sigma)x(\sigma)d\sigma, \qquad (4.2.14)$$

where $r \neq 0$ and $K(\sigma)$ is continuous on $[0,\infty)$. (Note that in the above development we assume that our network is at rest at $t = 0$, i.e., currents and charges in capacitors are zero.)

Thus, our assumptions (1)-(3) imply that conditions (i)-(iii) in Theorem 4.2.1 are satisfied.

Next, let us build a network $\hat{n}$ from $\hat{n}'$ by inserting a nonlinear resistor into the branch b_1. To be more specific, let φ be a real-valued continuous function on R^1, and, using the notation of Theorem 4.2.1, let $Z^+ : L_2[0,T] \to L_2[0,T]$ be defined by

$$(Z^+x)(t) = \varphi(x(t)). \tag{4.2.15}$$

(Thus, we have $D^+ = L_2[0,T]$.)

Finally, we will assume that the function $\Psi : R^1 \to R^1$, defined by

$$\Psi(\xi) = \varphi(\xi) + r\xi, \tag{4.2.16}$$

satisfies the following condition:

(4) there exists an $\alpha > 0$ such that, for all $\xi_1, \xi_2 \in R^1$,

$$|\Psi(\xi_1)-\Psi(\xi_2)| \geq \alpha|\xi_1-\xi_2|. \tag{4.2.17}$$

We are going to show that, under the assumptions (1)-(4), the network $\hat{n} = (\hat{Z},G)$, with $\hat{Z}$ being defined as in Theorem 4.2.1, i.e., $\hat{Z}x = \hat{Z}'x + (Z^+(x)_1)'$, is regular on D', and $Q(D') = H^{c2}$. (Note that we have $D' \cap D'' = D'$.)

To this end, observe first that, due to continuity of Ψ and (4.2.17), Ψ is 1-to-1 and maps R^1 onto R^1. Thus, if Ψ^{-1} is the inverse of ψ, we get readily from (4.2.17),

$$|\Psi^{-1}(\eta_1) - \Psi^{-1}(\eta_2)| \leq \alpha^{-1}|\eta_1-\eta_2| \tag{4.2.18}$$

for all $\eta_1, \eta_2 \in R^1$.

Referring to Theorem 4.2.1, consider the operator $(R+Z^+) : L_2[0,T] \to L_2[0,$ By (4.2.14), 4.2.15) and (4.2.16) we have

$$((R+Z^+)x)(t) = \Psi(x(t)) + \int_0^t K(t-\sigma)x(\sigma)d\sigma. \tag{4.2.19}$$

Choose arbitrarily $y \in L_2[0,T]$ and consider the equation $(R+Z^+)x = y$. By the above and (4.2.19), this equation is equivalent to

$$x(t) = \Psi^{-1}(-\int_0^t K(t-\sigma)x(\sigma)d\sigma + y(t)). \tag{4.2.20}$$

Define the operator S on $L_2[0,\tau]$ by

$$(Sx)(t) = \Psi^{-1}(-\int_0^t K(t-\sigma)x(\sigma)d\sigma + y(t)). \tag{4.2.21}$$

Since Ψ^{-1} is continuous by (4.2.18), we see easily that S maps $L_2[0,\tau]$ into itself.

Now, we are going to show that, for some integer $n \geq 1$, S^n is a contraction on $L_2[0,\tau]$. Indeed, let $C = \sup_{\xi\in[0,\tau]} |K(\xi)|$. Then we have by (4.2.21) for $x_1,x_2 \in L_2[0,\tau]$ and $t \in [0,\tau]$, (see (4.2.18))

$$\begin{aligned} |Sx_1-Sx_2)(t)| &\leq \alpha^{-1}|\int_0^t K(t-\sigma)(x_2-x_1)(\sigma)d\sigma| \\ &\leq \alpha^{-1}C \int_0^t |x_1-x_2|(\sigma)d\sigma. \end{aligned} \tag{4.2.22}$$

Using the induction, we can easily confirm that, for any integer $n \geq 1$,

$$|(S^n x_1-S^n x_2)(t)| \leq \frac{(\alpha^{-1}C)^n}{(n-1)!} \int_0^t (t-\sigma)^{n-1}|x_1-x_2|(\sigma)d\sigma. \tag{4.2.23}$$

However, (4.2.23) yields

$$\begin{aligned} |(S^n x_1-S^n x_2)(t)| &\leq \frac{(\alpha^{-1}C)^n}{(n-1)!} t^{n-1}\sqrt{\int_0^t d\sigma}\sqrt{\int_0^t |x_1-x_2|^2 d\sigma} \\ &\leq \frac{(\alpha^{-1}C)^n \tau^{n-\frac{1}{2}}}{(n-1)!} \|x_1-x_2\|. \end{aligned}$$

Hence,

$$\|S^n x_1-S^n x_2\| \leq \lambda_n\|x_1-x_2\|, \tag{4.2.24}$$

where

$$\lambda_n = \frac{(\alpha^{-1}C\tau)^n}{(n-1)!}. \tag{4.2.25}$$

From (4.2.25) it follows that $\lambda_n < 1$ for n sufficiently large, i.e., S^n is a contraction for such n. Hence, there exists a unique $x \in L_2[0,\tau]$ such that $x = Sx$, i.e., (4.2.20) holds; consequently, we have $(R+Z^+)x = y$. Thus, $R+Z^+$ is 1-to-1 on $L_2[0,\tau]$ and maps it onto itself. This means that, by (a) and (b) in Theorem 4.2.1 the network $\hat{\eta}$ is regular on D' and $Q(D') = H^{c_2}$, what we wanted to show.

Example 2. Let $\hat{\eta}'$ be exactly the same network as in Example 1, and assume that conditions (1)-(3) are satisfied. Now, we will consider a network $\hat{\eta}$ obtained from $\hat{\eta}'$ by inserting a nonlinear inductance in the branch b_1.

In more detail, let Φ be a real-valued function on R^1 which possesses a continuous derivative Φ' everywhere and satisfies the condition $\Phi(0) = 0$.

Using the notation of Theorem 4.2.1 define the operator $Z^+ : D^+ = \mathcal{K} \to L_2[0,\tau]$ by

$$(Z^+x)(t) = [\Phi(x(t))]'. \tag{4.2.26}$$

It is clear that this definition is meaningful, and Z^+ truly maps $\mathcal{K}$ into $L_2[0,\tau]$.

Furthermore, we will assume that

(4)* there exists $\alpha > 0$ such that $|\Phi'(\xi)| \geq \alpha$ for all $\xi \in R^1$.

We are going to show that, under conditions (1)-(3) and (4)*, the network $\hat{\mathfrak{n}} = (\hat{Z},G)$ is regular on $D' \cap D''$ and $Q(D' \cap D'') = H^{c_2}$. Here $D' \cap D'' = \mathcal{K} \times M_2 \times \ldots \times M_{c_2}$, where $M_2, M_3, \ldots, M_{c_2}$ are the same as in Example 1, and $\hat{Z}x = \hat{Z}'x + Z^+(x)_1)'$ for every $x \in D' \cap D''$.

To prove this, note first that (4)* and the mean value theorem imply that

$$|\Phi(\xi_1)-\Phi(\xi_2)| \geq \alpha|\xi_1-\xi_2| \tag{4.2.27}$$

for all $\xi_1, \xi_2 \in R^1$. Since Φ is continuous, it is a 1-to-1 mapping from R^1 onto R^1, and for the inverse Φ^{-1} we have

$$|\Phi^{-1}(\eta_1)-\Phi^{-1}(\eta_2)| \leq \alpha^{-1}|\eta_1-\eta_2| \tag{4.2.28}$$

for all $\eta_1, \eta_2 \in R^1$; also $\Phi^{-1}(0) = 0$.

Consider now the operator $(R+Z^+) : D_1 \cap D^+ = \mathcal{K} \to L_2[0,\tau]$. By (4.2.14) and (4.2.26) we have

$$((R+Z^+)x)(t) = [\Phi(x(t))]' + rx(t) + \int_0^t K(t-\sigma)x(\sigma)d\sigma. \tag{4.2.29}$$

Next, choose arbitrarily $y \in L_2[0,\tau]$, and consider the following two equations:

$$[\Phi(x(t))]' + rx(t) + \int_0^t K(t-\sigma)x(\sigma)d\sigma = y(t), \tag{*}$$

$$\Phi(x(t)) + r\int_0^t x(\sigma)d\sigma + \int_0^t \left(\int_0^\omega K(\omega-\sigma)x(\sigma)d\sigma\right)d\omega = \int_0^t y(\sigma)d\sigma. \tag{$\overset{*}{*}$}$$

Now, if there exists $x \in \mathcal{K}$ which satisfies (*), then x satisfies ($\overset{*}{*}$), too; for seeing this it suffices to realize that $\Phi(x(0)) = \Phi(0) = 0$.

Conversely, let $x \in L_2[0,\tau]$ be an element satisfying ($\overset{*}{*}$); then $x \in \mathcal{K}$ and satisfies (*).

Indeed, since Φ is 1-to-1 from R^1 onto R^1, equation ($\overset{*}{*}$) is equivalent to

$$x(t) = \Phi^{-1}[-r\int_0^t x(\sigma)d\sigma - \int_0^t(\int_0^{\omega}K(\omega-\sigma)x(\sigma)d\sigma)d\omega \qquad (+)$$
$$+ \int_0^t y(\sigma)d\sigma].$$

However, all terms in [...] are absolutely continuous; since Φ^{-1} has a continuous derivative $(\Phi^{-1})'$ and, by (4)*, $|(\Phi^{-1})'(\xi)| \leq \alpha^{-1}$ for all $\xi \in R^1$, it follows that $\Phi^{-1}[...]$ is absolutely continuous, and consequently, so is $x(t)$. Moreover, differentiating (+) we get

$$x'(t) = (\Phi^{-1})'[...]\cdot\{-rx(t) - \int_0^t K(t-\sigma)x(\sigma)d\sigma + y(t)\}. \quad (\ddagger)$$

Since $(\Phi^{-1})[...]$ is continuous, ($\ddagger$) shows that $x'(t) \in L_2[0,\tau]$. Finally, putting $t = 0$ into (+), we get $x(0) = \Phi^{-1}[0] = 0$; hence, $x \in \mathcal{K}$.

To conclude the proof of our assertion, it suffices to realize that (*) follows from ($\overset{*}{*}$) by differentiation.

Next, define the operator $S : L_2[0,\tau] \to L_2[0,\tau]$ by

$$(Sx)(t) = \Phi^{-1}(-\int_0^t h(t-\sigma)x(\sigma)d\sigma + \int_0^t y(\sigma)d\sigma), \qquad (4.2.30)$$

where

$$h(t-\sigma) = r + \int_\sigma^t K(\xi-\sigma)d\xi.$$

Then it is clear that equation $x = Sx$ is equivalent to (+).

On the other hand, since h is continuous and Φ^{-1} satisfies (4.2.28), it follows in the same way as in Example 1 that S^n is a contraction for n sufficiently large (after all, our S is practically the same as operator S in Example 1, see (4.2.21)). Thus for the chosen $y \in L_2[0,\tau]$, there exists a unique $x \in L_2[0,\tau]$ such that $x = Sx \Rightarrow$ (+) holds $\Rightarrow$ ($\overset{*}{*}$) holds $\Rightarrow x \in \mathcal{K}$ and (*) holds $\Rightarrow (R+Z^+)x = y$.

Hence, the operator $(R+Z^+)$ is 1-to-1 and maps $\mathcal{K}$ onto $L_2[0,\tau]$. Thus by Theorem 4.2.1, our network $\hat{n}$ is regular on $D'\cap D''$ and $Q(D'\cap D'') = H^{c_2}$, what we wanted to show.

Let us mention the fact that the above examples can be extended to the case where $\hat{n}'$ contains time-varying elements; the case of constant elements, however, is easier to analyze because the operators A_{ik} (see (4.1.13)) can be readily established by using methods of classical network analysis.

<u>Example 3</u>. Let again G be a finite oriented graph having branches $b_1, b_2, \ldots, b_{c_2}$, and let $\hat{n} = (\hat{Z}', G)$ be a DC-network built up from constant

resistors, i.e., $\hat{Z}'$ is described by a matrix $\tilde{R} = \mathrm{diag}(r_1, r_2, \ldots, r_{c_2})$, where the r_j's are real (not necessarily positive) numbers. For H we take the real line R^1.

We will assume that:

(1) The set of branches $\{b_1, b_2, \ldots, b_n\}$, $n < c_2$, is regular.

(2) The smallest eigenvalue λ of the $c_o \times c_o$-matrix $X^T \cdot \tilde{R} \cdot X$ is positive. (Here we take X real; clearly, λ is independent of the choice of X. Also note that $\lambda \leq \min_{1 \leq k \leq c_2} r_k$).

Now, we will construct $\hat{\Pi}$ by inserting nonlinear resistors into branches $b_1, b_2, \ldots, b_n$; note that mutual couplings may exist between these nonlinear resistors. In more detail, we let $Z^+ : R^n \to R^n$ be defined by

$$Z^+ x = \varphi(x), \tag{4.2.31}$$

where $\varphi : R^n \to R^n$ is a continuous function.

Finally, we will assume that

(3) there exists $\mu < \lambda$ such that

$$(\varphi(\xi_1) - \varphi(\xi_2))^T \cdot (\xi_1 - \xi_2) \geq -\mu \|\xi_1 - \xi_2\|^2 \tag{4.2.32}$$

for all $\xi_1, \xi_2 \in R^n$.

We are going to show that, under assumptions (1)-(3), $\hat{\Pi}$ is regular on R^{c_2} and $Q(R^{c_2}) = R^{c_2}$.

Indeed, since the matrix $X^T \cdot \tilde{R} \cdot X$ is positive definite by (2), the operator $W = \hat{X}^* \hat{Z}' \hat{X}$ is a bijection between R^{c_o} and itself; thus, by Theorems 2.2.2 and 2.2.3 it follows that $\hat{\Pi}'$ is regular on R^{c_2} and $Q(R^{c_2}) = R^{c_2}$. Hence, the condition (i) in Theorem 4.2.1 is met.

Moreover, (2) implies readily that

$$\langle Wx, x \rangle_{c_o} \geq \lambda \|x\|^2_{c_o} \quad \text{and} \quad \|Wx\|_{c_o} \leq d \|x\|_{c_o}$$

for all $x \in R^{c_o}$ and some $d > 0$; hence, by (1) and Theorem 4.1.2, the DPSI $R_o : R^n \to R^n$ of $b_1, b_2, \ldots, b_n$ is a continuous operator, and we have by (4.1.14),

$$\langle R_o j_1 - R_o j_2, j_1 - j_2 \rangle_n \geq \lambda \|j_1 - j_2\|^2_n \tag{4.2.33}$$

for all $j_1, j_2 \in R^n$.

On the other hand, (3) yields $\langle Z^+ j_1 - Z^+ j_2, j_1 - j_2 \rangle_n \geq -\mu \|j_1 - j_2\|^2_n$, and consequently, by (4.2.33),

$$\langle (R_o+Z^+)j_1-(R_o+Z^+)j_2, j_1-j_2\rangle_n \geq (\lambda-\mu)\|j_1-j_2\|_n^2 \tag{4.2.34}$$

for all $j_1, j_2 \in R^n$.

However, (4.2.34) shows that R_o+Z^+ is 1-to-1 on R^n, and that R_o+Z^+ is a coercive, maximal monotone operator because it is continuous; hence, by Lemma 1.1.5, $(R_o+Z^+)R^n = R^n$. The Theorem 4.2.1 completes the proof of our claim.

As for the existence of a solution of $\hat{\eta}$, our conditions (1)-(3) can be modified. Indeed, assume that the following requirements (1)* and (2)* are met:

(1)* The network $\hat{\eta}'$ is regular on R^{c_2} and the DPSI R_o of branches $b_1, b_2, \ldots, b_n$ is an operator.

(2)* There exist constants $q > 0$ and $M > 0$ such that

$$\|\varphi(\xi)\| \leq q\|\xi\| \tag{4.3.35}$$

for all $\xi \in R^n$ with $\|\xi\| > M$.

We are going to show that, if q is sufficiently small, the network $\hat{\eta}$ has the property that $Q(R^{c_2}) = R^{c_2}$.

To this end, observe first that, due to Theorems 2.2.2, 2.2.3 and finite dimensionality of R^{c_o}, we have $Q(R^{c_2}) = R^{c_2}$ for $\hat{\eta}'$. Moreover, since R_o is a homeomorphism between R^n and itself (witness (4.1.3)), it follows that $D_n = R^n$, and consequently, all conditions (i)-(iii) in Theorem 4.2.1 are satisfied. Also, there exists $\mu > 0$ such that

$$\mu\|x\|_n \leq \|R_o x\|_n \tag{4.2.36}$$

for all $x \in R^n$.

On the other hand, (4.2.31) and (4.2.35) yield $\|Z^+x\|_n \leq q\|x\|_n$ for all $x \in R^n$ with $\|x\|_n > M$; thus, for any such x, we have by (4.2.36),

$$\|Z^+x\|_n \leq q\mu^{-1}\|R_o x\|_n. \tag{4.2.37}$$

Now, if q is so small that $q\mu^{-1} < 1$, then the Sandberg-Willson's theorem [12] shows that $(R_o+Z^+)R^n = R^n$; hence, Theorem 4.2.1 concludes the proof of our claim.

4.3 Shannon-Hagelbarger Theorem

The classical Shannon-Hagelbarger theorem claims (see [10], [11]) that the driving point impedance of a fixed branch of a DC-network is a concave nondecreasing function of the matrix $\mathrm{diag}(r_1,r_2,\ldots,r_{c_2})$ describing the network resistances. In this section we will generalize this result to the case in which the DPSI of a Hilbert network is considered. To this end, we introduce the following notation:

Let $\mathcal{H}$ be a Hilbert space, and let $M \in [\mathcal{H},\mathcal{H}]$ be a selfadjoint operator; M will be called positive, if $\langle Mx,x\rangle$ is real and nonnegative for each $x \in \mathcal{H}$. In this case we will write $M \geq 0$. If $M_1-M_2 \geq 0$, we will also write $M_1 \geq M_2$ or $M_2 \leq M_1$.

Let $c \leq \aleph_0$, and let $\mathfrak{G}_c$ be the set of all selfadjoint operators in $[H^c,H^c]$ having the property that, for each $\hat{Z} \in \mathfrak{G}_c$, $\hat{Z}-\gamma_{\hat{Z}}I \geq 0$ for some $\gamma_{\hat{Z}} > 0$.

We can readily verify that $\mathfrak{G}_c$ is a convex subset of $[H^c,H^c]$. Indeed, let $\hat{Z}_1,\hat{Z}_2 \in \mathfrak{G}_c$ and $\lambda \in [0,1]$; then, for any $x \in H^c$, we have $\langle(\lambda\hat{Z}_1+(1-\lambda)\hat{Z}_2)x,x\rangle_c = \lambda\langle\hat{Z}_1x,x\rangle_c + (1-\lambda)\langle\hat{Z}_2x,x\rangle_c \geq \lambda\gamma_{\hat{Z}_1}\|x\|_c^2 + (1-\lambda)\gamma_{\hat{Z}_2}\|x\|_c^2$ $\geq \min[\gamma_{\hat{Z}_1},\gamma_{\hat{Z}_2}]\cdot\|x\|_c^2$; since $\lambda\hat{Z}_1 + (1-\lambda)\hat{Z}_2$ is again selfadjoint, our claim is proved.

Now, let G be a locally finite oriented graph with the set of branches $\mathcal{B} = \{b_1,b_2,\ldots\}$ having cardinal $c_2 \leq \aleph_0$, and assume that the set of branches $\{b_1,b_2,\ldots,b_n\}$ is regular. By $R(\hat{Z})$ we will denote the DPSI of branches $b_1,b_2,\ldots,b_n$ provided the network $(\hat{Z},G)$ is under consideration.

Theorem 4.3.1. Let H be a real Hilbert space, let G be a locally finite oriented graph and let the set of branches $\{b_1,b_2,\ldots,b_n\}$ be regular. Then, for each $\hat{Z} \in \mathfrak{G}_{c_2}$, the network $\hat{\mathfrak{n}} = (\hat{Z},G)$ is regular on H^{c_2}, $Q(H^{c_2}) = H^{c_2}$ and the DPSI $R(\hat{Z})$ of branches $b_1,b_2,\ldots,b_n$ is a positive operator. Moreover, $R(\hat{Z})$ is a concave nondecreasing function on $\mathfrak{G}_{c_2}$, i.e.,

$$R(\lambda\hat{Z}_1 + (1-\lambda)\hat{Z}_2) \geq \lambda R(\hat{Z}_1) + (1-\lambda)R(\hat{Z}_2) \qquad (4.3.1)$$

for all $\lambda \in [0,1]$ and all $\hat{Z}_1,\hat{Z}_2 \in \mathfrak{G}_{c_2}$, and $R(\hat{Z}_1) \leq R(\hat{Z}_2)$ whenever $\hat{Z}_1 \leq \hat{Z}_2$, $\hat{Z}_1,\hat{Z}_2 \in \mathfrak{G}_{c_2}$.

Proof: Recalling the definition of $\mathfrak{G}_c$ it follows that, for any $\hat{Z} \in \mathfrak{G}_{c_2}$ and $x \in H^{c_o}$, $\langle Wx,x\rangle_{c_o} = \langle\hat{X}^*\hat{Z}\hat{X}x,x\rangle_{c_o} = \langle\hat{Z}\hat{X}x,\hat{X}x\rangle_{c_2} \geq \gamma_{\hat{Z}}\|\hat{X}x\|_{c_2}^2 = \gamma_{\hat{Z}}\|x\|_{c_o}^2$; hence, by Theorem 2.3.4, the network $(\hat{Z},G)$ is regular on H^{c_2} and $Q(H^{c_2}) = H^{c_2}$. Moreover, the established inequality shows that the assumptions of Theorem 4.1.2 are met, and consequently, $R(\hat{Z})$ exists and is in $[H^n,H^n]$. Also, from Theorems 2.2.2 and 2.2.3, and from (4.1.3) we conclude easily that $R(\hat{Z})$ is selfadjoint.

Next, as in Section 4.1, let $(x)_n = [x_1, x_2, \ldots, x_n]^T \in H^n$ whenever $x = [x_k] \in H^{c_2}$; also, put $s = (1,2,\ldots,n)$ and, as in Section 2.5, let ${}^{c(s)}N_{\hat{a}}$ be the space of all vectors $x \in N_{\hat{a}}$ with $(x)_n = 0$.

Now, choose arbitrarily $\hat{Z}_1, \hat{Z}_2 \in \mathfrak{G}_{c_2}$, $\lambda \in [0,1]$, $j \in H^n$ and denote $e_\lambda = [(R(\lambda\hat{Z}_1 + (1-\lambda)\hat{Z}_2)j)^T, 0, 0, \ldots]^T \in H^{c_2}$, and $e_m = [(R(\hat{Z}_m)j)^T, 0, 0, \ldots]^T \in H^{c_2}$ for $m = 1,2$. Let $i^\lambda = [i_k^\lambda] \in H^{c_2}$ be the solution of $\hat{n}_\lambda = (\lambda\hat{Z}_1 + (1-\lambda)\hat{Z}_2, G)$ corresponding to e_λ, and $i^m = [i_k^m] \in H^{c_2}$ be the solution of $\hat{n}_m = (\hat{Z}_m, G)$ corresponding to e_m, $m = 1,2$. Thus, we have by the properties of the DPSI, $(i^\lambda)_n = (i^1)_n = (i^2)_n = j$. Moreover,

$$\begin{aligned} \langle i^\lambda, e_\lambda\rangle_{c_2} &= \langle j, R(\lambda\hat{Z}_1 + (1-\lambda)\hat{Z}_2)j\rangle \\ &= \langle i^\lambda, (\lambda\hat{Z}_1 + (1-\lambda)\hat{Z}_2)i^\lambda\rangle_{c_2} \\ &= \lambda\langle i^\lambda, \hat{Z}_1 i^\lambda\rangle_{c_2} + (1-\lambda)\langle i^\lambda, \hat{Z}_2 i^\lambda\rangle_{c_2}. \end{aligned} \tag{4.3.2}$$

Also,

$$\langle i^m, e_m\rangle_{c_2} = \langle j, R(\hat{Z}_m)j\rangle_n = \langle i^m, \hat{Z}_m i^m\rangle_{c_2} \tag{4.3.3}$$

$m = 1,2$. The relation (4.3.3) shows that $R(\hat{Z})$ is a positive operator for any choice of $\hat{Z} \in \mathfrak{G}_{c_2}$. On the other hand, $i^\lambda - i^1 \in {}^{c(s)}N_{\hat{a}}$, i.e., $i^\lambda \in i^1 + {}^{c(s)}N_{\hat{a}}$; hence, by Theorem 2.5.3 (its assumptions are clearly met), $\langle i^\lambda, \hat{Z}_1 i^\lambda\rangle_{c_2} \geq \langle i^1, \hat{Z}_1 i^1\rangle_{c_2}$. By the same argument, $i^\lambda \in i^2 + {}^{c(s)}N_{\hat{a}}$, and consequently, $\langle i^\lambda, \hat{Z}_2 i^\lambda\rangle_{c_2} \geq \langle i^2, \hat{Z}_2 i^2\rangle_{c_2}$. Thus, using these inequalities and relations (4.3.2) and (4.3.3), we obtain

$$\langle j, R(\lambda\hat{Z}_1 + (1-\lambda)\hat{Z}_2)j\rangle_n \geq \langle j, \lambda R(\hat{Z}_1)j\rangle_n + \langle j, (1-\lambda)R(\hat{Z}_2)j\rangle_n;$$

this, however, is exactly (4.3.1) since $j \in H^n$ was arbitrary.

Finally, assume that $\hat{Z}_1 \leq \hat{Z}_2$, $\hat{Z}_1, \hat{Z}_2 \in \mathfrak{G}_{c_2}$. Choosing $j \in H^n$ and using the same notation as above, we have equalities (4.3.3). Since $i^2 \in i^1 + {}^{c(s)}N_{\hat{a}}$, it follows that $\langle i^2, \hat{Z}_1 i^2\rangle_{c_2} \geq \langle i^1, \hat{Z}_1 i^1\rangle_{c_2}$, and consequently, $\langle i^2, \hat{Z}_2 i^2\rangle_{c_2} \geq \langle i^1, \hat{Z}_1 i^1\rangle_{c_2}$. Hence, by (4.3.3), $\langle j, R(\hat{Z}_2)j\rangle_n \geq \langle j, R(\hat{Z}_1)j\rangle_n$, which concludes the proof.

Problems

4.3.1 Give an example of a H.n. such that $F \neq \emptyset$ but $Q_1 = \emptyset$.

4.3.2 If, for a H.n. $\hat{n} = (\hat{Z}, G)$, $\hat{Z}$ is a linear bounded operator, then the assumptions made in Theorem 4.1.1 can be relaxed. Indeed, prove that the following proposition is true:

Let $\hat{n} = (\hat{Z},G)$ be a H.n. such that $\hat{Z} \in [H^{c_2},H^{c_2}]$ and

$$|\langle Wx,x\rangle_{c_o}| \geq \gamma\|x\|^2_{c_o}$$

for some $\gamma > 0$ and all $x \in H^{c_o}$, where $W = \hat{X}^*\hat{Z}\hat{X}$. Furthermore, assume that the set of branches $\{b_1,b_2,\ldots,b_n\}$ is regular. Then the DPSI R of $b_1,b_2,\ldots$ $\ldots,b_n$ is an operator in $[H^n,H^n]$.

Using this result show that the assumption "H is real" made in Theorem 4.3.1 can be dropped.

4.3.3 Let $\hat{n} = (\hat{Z},G)$ be a H.n. with $F \neq \emptyset$ and $Q_n \neq \emptyset$; for each $j \in D_n$, let $Mj = \{e : e \in Q_n$, there exists a solution i_e of $\hat{n}$ corresponding to $(e)'$ such that $(i_e)_n = j\}$. Then M is the DPSI R of branches $b_1,b_2,\ldots,b_n$.

4.4 Driving Point Impedance of Some Elementary Networks

In this section we will establish the driving point impedance of some networks that have a simple structure. These results generalize the classical rules on the impedance of series- and parallel-connections.

Theorem 4.4.1. Let $D_o \subset H$ be nonempty, let $Z_j : D_o \to \mathfrak{S}(H)$ be a set mapping for $j = 1,2,\ldots,n$, and let the set mapping $\hat{Z} : H \times D_o^n \to \mathfrak{S}(H^{n+1})$ be defined by $\hat{Z}[x_1,x_2,\ldots,x_{n+1}]^T = [0,Z_1x_2,Z_2x_3,\ldots,Z_nx_{n+1}]^T$; moreover, let $\hat{n} = (\hat{Z},G)$ be a H.n. whose graph G is given in Figure 4.4.1. Then $Q_1 = (RD_o)^o$ with $R = \sum_{j=1}^{n} Z_j$, $D_1 = D_o$, and the driving point impedance of the branch b_1 is equal to $R : D_o \to \mathfrak{S}(H)$.

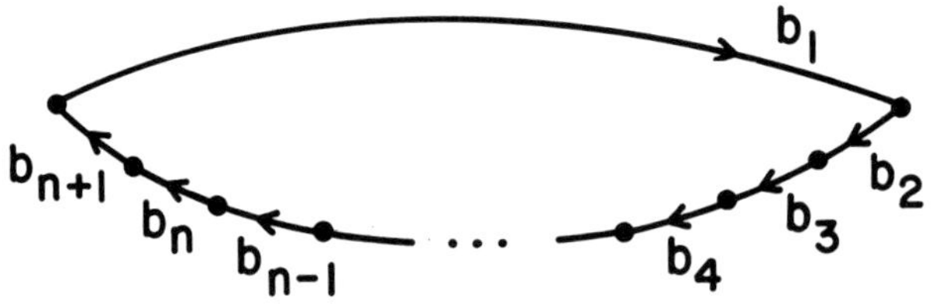

Figure 4.4.1

Proof: First of all, due to the particular form of the graph G every $\gamma \in N_{\tilde{a}}$ (see Section 2.1) is a scalar multiple of the n+1-vector $\gamma_o = [1,1,1,\ldots,1]^T$; also, since $\hat{n}$ is finite, we can put $X = \gamma_o$ (see Problem 2.2.1) in the formula defining the driving point impedance.

Now, we are going to show that $Q_1 = (RD_o)^o$. Indeed, let $e_1 \in Q_1$; then $\hat{\eta}$ possesses a solution, say $i = [i_1, i_2, \ldots, i_{n+1}]^T$ corresponding to $e_1' = [e_1, 0, \ldots, 0]^T$. Thus, we have $i_2, i_3, \ldots, i_{n+1} \in D_o$, and by K_2^+,

$$i_1 = i_2 = \ldots = i_{n+1}. \tag{4.4.1}$$

Moreover, by K_1^+ (see (2.2.2)), there exist elements $v_k \in Z_{k-1} i_k$, $k = 2,3,\ldots$ $\ldots, n+1$ such that

$$\sum_{k=2}^{n+1} v_k = e_1. \tag{4.4.2}$$

Hence,

$$e_1 \in \sum_{k=2}^{n+1} Z_{k-1} i_k = \sum_{k=2}^{n+1} Z_{k-1} i_1 = R i_1 \subset (RD_o)^o; \tag{4.4.3}$$

consequently,

$$Q_1 \subset (RD_o)^o.$$

Conversely, let $e_1 \in (RD_o)^o$; then there exists $i_1 \in D_o$ such that $e_1 \in Ri_1$ and consequently, there exist elements $v_2, v_2, \ldots, v_{n+1}$ such that $v_k \in Z_{k-1} i_1$ for $k = 2,3,\ldots,n+1$ and (4.4.2) holds. Thus, defining $i_2, i_3, \ldots, i_{n+1}$ by (4.4.1), the vector $[i_1, i_2, \ldots, i_{n+1}]^T$ will be a solution corresponding to e_1'. Hence, $e_1 \in Q_1$, i.e., $(RD_o)^o \subset Q_1$.

Furthermore, for any $e_1 \in Q_1$ we always have $i_1 \in D_o$. Conversely, choose $i_1 \in D_o$ arbitrarily, and pick some $e_1 \in Ri_1$; then $e_1 \in (RD_o)^o = Q_1$, (4.4.3) holds, and consequently, $[i_1, i_1, \ldots, i_1]^T$ is a solution corresponding to e_1'. Hence, by our definition (4.1.3), $D_1 = D_o$.

Finally, if $u \in D_o$, we get readily $Wu = \hat{X}^* \hat{Z} \hat{X} u = [1,1,\ldots,1] \cdot$ $[0, Z_1 u, Z_2 u, \ldots, Z_n u]^T = Ru$; thus, for $e_1 \in Q_1$, $Ae_1' = \hat{X} W^- \hat{X}^* e_1' = \hat{X} R^- \hat{X}^* e_1' = \hat{X} R^- e_1$. Hence, by (4.1.4), $A' e_1 = (\hat{X} R^- e_1)_1 = R^- e$; consequently, the driving point impedance of b_1 is $(A')^- = (R^-)^- = R$, which completes the proof.

<u>Remark 1</u>. The above theorem furnishes another interpretation of the concept of the driving point impedance. Indeed, consider the simple network given in Figure 4.4.2. Here, only the branch b_2 contains an element described by a set mapping Z. By our theorem, the driving point impedance of b_1 equals Z. In other words, the mapping Z characterizing the element in terms of current and voltage drop appears as the impedance of the simple circuit considered.

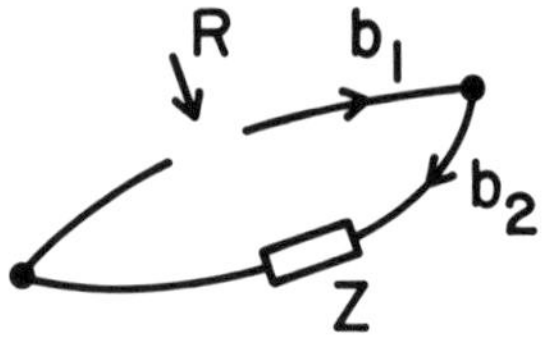

Figure 4.4.2

Theorem 4.4.2. Let $D_o^j \subset H$ be nonempty, let $Z_j : D_o^j \to \mathfrak{S}(H)$ be a set mapping for $j = 1,2,\ldots,n$, and let the set mapping $\hat{Z} : H \times D_o^1 \times D_o^2 \times\ldots\times D_o^n \to \mathfrak{S}(H^{n+1})$ be defined by $\hat{Z}[x_1,x_2,\ldots,x_{n+1}]^T = [0,Z_1x_2,Z_2x_3,\ldots,Z_nx_{n+1}]^T$; moreover, let $\hat{n} = (\hat{Z},G)$ be a H.n. whose graph is given in Figure 4.4.3. If $\bigcap_{j=1}^{n}(Z_jD_o^j)^o \neq \emptyset$, then

$$Q_1 = \bigcap_{j=1}^{n}(Z_jD_o^j)^o, \tag{4.4.4}$$

$$D_1 = \{(\sum_{j=1}^{n} Z_j^-)Q_1\}^o, \tag{4.4.5}$$

and the driving point impedance $R : D_1 \to \mathfrak{S}(H)$ of the branch b_1 is given by

$$R = (\sum_{j=1}^{n} Z_j^-)^-. \tag{4.4.6}$$

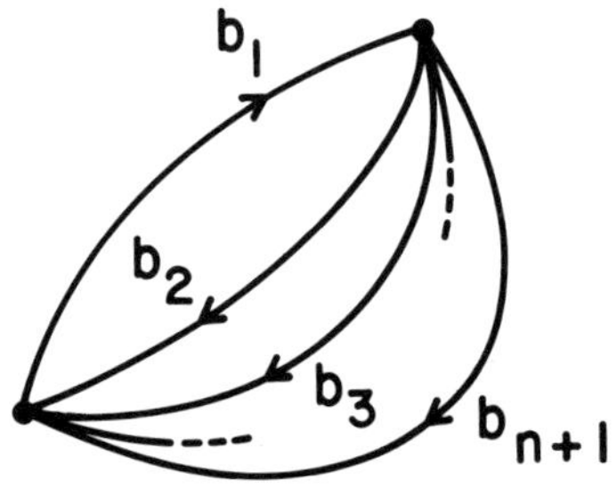

Figure 4.4.3

Proof: First, from Figure 4.4.3 it follows that the n+1-vectors $\gamma^1 = [1,1,0,0,\ldots,0]^T$, $\gamma^2 = [1,0,1,0,\ldots,0]^T$, $\gamma^3 = [1,0,0,1,0,\ldots,0],\ldots$ $\gamma^n = [1,0,0,\ldots,0,1]^T$ constitute a basis in $N_{\tilde{a}}$. Since $\hat{n}$ is finite, we can set $X = [\gamma^1 \vdots \gamma^2 \vdots \ldots \vdots \gamma^n]$.

In order to prove the relation (4.4.4), choose $e_1 \in Q_1$. Then $\hat{n}$

possesses a solution, say $i = [i_1, i_2, \ldots, i_{n+1}]^T$, corresponding to $e_1' = [e_1, 0, 0, \ldots, 0]^T$, and we have $i_k \in D_o^{k-1}$ for $k = 2,3,\ldots,n+1$; also, by K_2^+,

$$i_1 = \sum_{k=2}^{n+1} i_k. \tag{4.4.7}$$

Moreover, applying K_1^+ for vectors $\gamma^1, \gamma^2, \ldots, \gamma^n$, it follows that there exist elements $v_k \in Z_{k-1} i_k$, $k = 2,3,\ldots,n+1$ such that

$$v_k = e_1 \tag{4.4.8}$$

for $k = 2,3,\ldots,n+1$. Thus, $e_1 \in Z_{k-1} i_k \subset (Z_{k-1} D_o^{k-1})$ for $k = 2,3,\ldots,n+1$; consequently, $e_1 \in M$ with

$$M = \bigcap_{j=1}^{n} (Z_j D_o^j)^o, \tag{4.4.9}$$

so that $Q_1 \subset M$.

Also, from $e_1 \in Z_{k-1} i_k$ we infer that $i_k \in Z_{k-1}^- e_1$ for $k = 2,3,\ldots,n+1$, so that, by (4.4.7),

$$i_1 \in (\sum_{j=1}^{n} Z_j^-) e_1 \subset \{(\sum_{j=1}^{n} Z_j^-) Q_1\}^o. \tag{4.4.10}$$

Conversely, let $e_1 \in M$. Then, by (4.4.9), there exist elements $i_k \in D_o^{k-1}$ such that $e_1 \in Z_{k-1} i_k$ for $k = 2,3,\ldots,n+1$. Putting $v_k = e_1$ for $k = 2,3,\ldots,n+1$ and defining i_1 by (4.4.7), we see readily that $[i_1, i_2, \ldots, i_{n+1}]^T$ is a solution of $\hat{\Pi}$ corresponding to e_1'; hence, $e_1 \in Q_1$, i.e., $M \subset Q_1$ which proves (4.4.4).

Moreover, (4.4.10) shows that $D_1 \subset \{(\sum_{j=1}^{n} Z_j^-) Q_1\}^o = S$. To prove the converse inclusion, choose $i_1 \in S$. Then there exists $e_1 \in Q_1$ such that $i_1 \in (\sum_{j=1}^{n} Z_j^-) e_1 = \sum_{j=1}^{n} Z_j^- e_1$; consequently, there exists $i_k \in Z_{k-1}^- e_1$, $k = 2,3,\ldots,n+1$ such that (4.4.7) holds. However, since $e_1 \in Q_1 \subset (Z_{k-1} D_o^{k-1})^o$ for each $k = 2,3,\ldots,n+1$, it follows that $i_k \in D_o^{k-1}$ for $k = 2,3,\ldots,n+1$ and $e_1 \in Z_{k-1} i_k$. Thus, by the above, $[i_1, i_2, \ldots i_{n+1}]^T$ is a solution corresponding to e_1'. Hence, $S \subset D_1$ and (4.4.5) is proven.

To conclude the proof, let $u = [u_1, u_2, \ldots, u_n]^T \in F$. Then $\hat{X}u = [\sum_{k=1}^{n} u_k, u_1, u_2, \ldots, u_n]^T$ and consequently, $\hat{Z}\hat{X}u = [0, Z_1 u_1, Z_2 u_2, \ldots, Z_n u_n]^T$; hence , $Wu = \hat{X}^* \hat{Z} \hat{X} u = [Z_1 u_1, Z_2 u_2, \ldots, Z_n u_n]^T$. So, we have $W^- v = [Z_1^- v_1, Z_2^- v_2, \ldots$ $\ldots, Z_n^- v_n]^T$. Now, let $e_1 \in Q_1$; then $\hat{X}^* e_1' = [e_1, e_1, \ldots, e_1]^T$ so that $W^- \hat{X}^* e_1' = [Z_1^- e_1, Z_2^- e_1, \ldots, Z_n^- e_1]^T$. Hence, $Ae_1' = \hat{X} W^- \hat{X}^* e_1' = [\sum_{k=1}^{n} Z_k^- e_1, Z_1^- e_1, \ldots$ $\ldots Z_n^- e_1]^T$; thus, by (4.1.4), $A' e_1 = (\sum_{k=1}^{n} Z_k^-) e_1$. Consequently, the driving point impedance R of b_1 is given by $R = (\sum_{k=1}^{n} Z_k^-)^-$ what we wanted to show.

We conclude this section with the following remark: as we have seen, the concept of the driving point set impedance by no means depends on the fact that the underlying space is a Hilbert space. In other words, for defining the driving point impedance the essential thing is that "voltages" and "currents" are certain c_2-vectors, i.e., we know the first components of these entities. Also, as noted earlier, Theorem 4.2.1 holds without any change for algebraic networks considered in Chapter 6, i.e., finite networks whose underlying space if not equipped with any topology. Similarly, it is clear that Theorems 4.4.1 and 4.4.2 do not use any topological properties of the underlying space H; consequently, the same results hold for algebraic networks discussed in Chapter 6.

CHAPTER 5

SOME EXTENSIONS OF THE HILBERT NETWORK CONCEPT

5.1 Networks Having Nodes of Infinite Order

In the definition of a Hilbert network $\hat{n} = (\hat{Z},G)$ we have made the assumption that G is locally finite. This requirement, however, can easily be removed.

Indeed, let G be an oriented graph having the set of branches $\mathcal{B}$ and the set of nodes $\mathcal{V}$. For every $v_k \in \mathcal{V}$, define the subsets $\mathcal{B}_k^+$ and $\mathcal{B}_k^-$ of $\mathcal{B}$ as in Section 2.1, i.e.,

$$\mathcal{B}_k^+ = \{b_j : b_j \in \mathcal{B}, Gb_j = (v_\alpha,v_k)\},$$

$$\mathcal{B}_k^- = \{b_j : b_j \in \mathcal{B}, Gb_j = (v_k,v_\alpha)\},$$

and let

$$\mathcal{B}_k = \mathcal{B}_k^+ \cup \mathcal{B}_k^-. \tag{5.1.1}$$

As we know, $\mathcal{B}_k \neq \emptyset$ for each $v_k \in \mathcal{V}$. Thus, for each $v_k \in \mathcal{V}$, let $c(v_k)$ be the cardinal of $\mathcal{B}_k$. Clearly, $c(v_k) \leq \aleph_o$. If $c(v_k) = \aleph_o$, the node v_k will be said to have an infinite order, in the opposite case, v_k will be said to have a finite order.

Let again the incidence matrix d of G be defined by (2.1.3). Observe that, because G is not assumed to be locally finite, d need not have the property that each column of d contains only finitely many nonzero elements.

Choose now a diagonal $c_1 \times c_1$ matrix K, $K = \operatorname{diag}(k_1,k_2,k_3,\ldots)$ as follows: If $v_i \in \mathcal{V}$ has an infinite order, let $k_i = 0$; if $v_i \in \mathcal{V}$ has a finite order, let $k_i \neq 0$ be so small that the $c_1 \times c_2$ matrix

$$a = K\cdot d^T = [a_{ik}] \tag{5.1.2}$$

has the property that $\sum_{ik} |a_{ik}|^2 < \infty$. It is clear that such K always exists, and $a \neq 0$ whenever G contains at least one node of finite order.

Defining then the mapping $\tilde{a} : R^{c_2} \to R^{c_1}$ by (2.1.5) and $\hat{a} : H^{c_2} \to H^{c_1}$

by (2.1.10), it is easy to see that the null spaces of $\tilde{a}$ and $\hat{a}$ are independent of the choice of K, since, instead of (2.1.6), we now have for $\xi \in R^{c_2}$,

$$\tilde{a}\xi = 0 \Leftrightarrow \sum_{\mu} d_{\mu k}\xi_{\mu} = 0 \qquad (5.1.3)$$

for every k such that v_k has a finite order.

Thus, the theory developed in Part 2 remains valid for the present situation, i.e., a Hilbert network $\hat{h} = (\hat{Z},G)$ and its solution can be defined in the same way as in Chapter 2.

Using physical terminology, for the present model we do not require that Kirchhoff's node law be satisfied at any node of infinite order; on the other hand, Kirchhoff's loop law is required to hold for every vector $\gamma \in N_{\tilde{a}}$, i.e., in particular for every loop.

5.2 Networks Having Current Sources

The theory developed in Parts 1-4 was motivated by the physical situation that the network in question is driven solely by voltage-sources. However, our results can easily be extended to the case that current sources are present as well.

To this end, let us first reconsider the abstract network, and extend the concept of a solution.

As in Section 1.2, let $\mathcal{L}$ be a nontrivial linear space, let M, N be linear subspaces of $\mathcal{L}$ such that $\mathcal{L} = M \oplus N$, and let $Z : D \to \mathfrak{S}(\mathcal{L})$, where $D \subset \mathcal{L}$, $D \neq \emptyset$.

Definition. Let $h = [Z,N,M]$ be an abstract network over $\mathcal{L}$, and let $(e,k) \in \mathcal{L} \times \mathcal{L}$; an element $i \in \mathcal{L}$ will be called a solution of h corresponding to the pair (e,k), if

(i) there exists $v \in Zi$ such that

$$v-e \in M, \qquad (5.2.1)$$

(ii) $i \in (k+N) \cap D$.

Observe that if $i \in \mathcal{L}$ is a solution of h corresponding to (e,k), then i is also a solution corresponding to (e,k+k') for any $k' \in N$, since $k+N = k+k'+N$.

Let us introduce the following notation:

If $\tilde{D} \subset D$, $\tilde{D} \neq \emptyset$, and if $k \in \mathcal{L}$, let

$$\tilde{D}_k = (k+N) \cap \tilde{D}. \qquad (5.2.2)$$

If $\tilde{D}_k \neq \phi$, let

$$Q(\tilde{D}_k) = M + (Z\tilde{D}_k)^{\circ} = P^{-1}[(PZ)\tilde{D}_k]^{\circ}. \qquad (5.2.3)$$

Furthermore, let

$$K(\tilde{D}) = \{k : k \in \mathcal{L}, \tilde{D}_k \neq \phi\}, \qquad (5.2.4)$$

and

$$R(\tilde{D}) = \{(x,y) : y \in K(\tilde{D}), x \in Q(\tilde{D}_y)\}. \qquad (5.2.5)$$

It is clear that $K(\tilde{D}) \neq \phi$ whenever $\tilde{D} \neq \phi$, and consequently, $R(\tilde{D}) \neq \phi$; also, $K(\tilde{D})$ is a linear flat in $\mathcal{L}$, i.e., if $k \in K(\tilde{D})$, then $k+k' \in K(\tilde{D})$ for each $k' \in N$.

Theorem 5.2.1. Let $\mathfrak{n} = [Z,N,M]$ be an abstract network over $\mathcal{L}$, let $\tilde{D} \subset D$, $\tilde{D} \neq \phi$, and let $(e,k) \in \mathcal{L} \times \mathcal{L}$; then $\mathfrak{n}$ has a solution $i \in \tilde{D}$ corresponding to (e,k), if and only if $(e,k) \in R(\tilde{D})$. In this case, the set I of all solutions in $\tilde{D}$ corresponding to (e,k) is given by

$$I = (PZ)_k^- Pe, \qquad (5.2.6)$$

where $(PZ)_k^- : [(PZ)\tilde{D}_k]^{\circ} \to \mathfrak{S}(\tilde{D}_k)$ is the quasiinverse of the set mapping $PZ : \tilde{D}_k \to \mathfrak{S}([(PZ)\tilde{D}_k]^{\circ})$ and P is the projection from $\mathcal{L}$ onto N along M.

Proof: (1) Let $i \in \tilde{D}$ be a solution corresponding to (e,k). Then, by (ii), $i \in (k+N)\cap D$, so that $i \in (k+N)\cap\tilde{D}$; hence, by (5.2.2) and (5.2.4), $k \in K(\tilde{D})$. Moreover, by (i), there exists $v \in Zi$ such that $v-e \in M$, i.e., $e \in v+M \subset Zi+M \subset M+(Z\tilde{D}_k)^{\circ} = Q(\tilde{D}_k)$. Hence, $(e,k) \in R(\tilde{D})$.

(2) Conversely, let $(e,k) \in R(\tilde{D})$; then $k \in K(\tilde{D})$, i.e., $\tilde{D}_k = (k+N)\cap\tilde{D} \neq \phi$. Moreover, $e \in Q(\tilde{D}_k) = M+(Z\tilde{D}_k)^{\circ}$, so that there exists $m \in M$ and $i \in \tilde{D}_k$ and $v \in Zi$ such that $e = m+v$, i.e., $v-e = -m \in M$, and $i \in (k+N)\cap\tilde{D} \subset (k+N)\cap D$. Hence, i is a solution of $\mathfrak{n}$ corresponding to (e,k) and $i \in \tilde{D}$.

In order to prove the relation (5.2.6), let I be the set of all solutions in $\tilde{D}$ which correspond to $(e,k) \in R(\tilde{D})$. Note also that the set $(PZ)_k^- Pe$ is well-defined. Indeed, since $e \in Q(\tilde{D}_k)$, it follows by (5.2.3) that $Pe \in [(PZ)\tilde{D}_k]^{\circ}$, i.e., Pe is in the domain of $(PZ)_k^-$.

(a) Let $i \in (PZ)_k^- Pe$. Since $(PZ)_k^-$ maps its domain into $\mathfrak{S}(\tilde{D}_k)$ we have $i \in \tilde{D}_k \subset (k+N)\cap D$, so that is satisfies (ii). Moreover, by (i) in Lemma 1.1.1, $Pe \in PZi$; thus, there exists $v \in Zi$ such that $Pe = Pv$. Consequently, $v-e \in M$, i.e., i satisfies (i). Hence, $i \in I$ which implies that $(PZ)_k^- Pe \subset I$.

(b) Conversely, let $i \in I$. Then, by (ii), $i \in (k+N) \cap D$, and since $i \in \tilde{D}$, $i \in \tilde{D}_k$. Moreover, by (i), there exists $v \in Zi$ such that $v-e \in M$. Hence, $Pe = Pv$, i.e., $Pe \in PZi$. Thus, by (i) in Lemma 1.1.1, $i \in (PZ)_k^- Pe$. Consequently, $I \subset (PZ)_k^- Pe$ which proves (5.2.6).

Next, we have the following proposition on uniqueness:

<u>Theorem 5.2.2</u>. Let $\mathfrak{h} = [Z,N,M]$ be an abstract network over $\mathcal{L}$, and let $\tilde{D} \subset D$, $\tilde{D} \neq \emptyset$; then for each $(e,k) \in R(\tilde{D})$ there exists in $\tilde{D}$ a unique solution i of $\mathfrak{h}$ corresponding to (e,k), if and only if, for every $k^* \in K(\tilde{D})$, the set mapping PZ is simple on $\tilde{D}_k$.

<u>Proof</u>: Choose arbitrarily $k^* \in K(\tilde{D})$, and let $(e,k^*) \in R(\tilde{D})$. Then we have by Theorem 5.2.1 and Lemma 1.1.2: $\mathfrak{h}$ possesses in $\tilde{D}$ a unique solution corresponding to $(e,k^*) \Leftrightarrow I = (PZ)_{k^*}^- Pe$ is a singleton $\Leftrightarrow (PZ)_{k^*}^-$ is an operator on $[(PZ)\tilde{D}_{k^*}]^o \Leftrightarrow$ PZ is simple on $\tilde{D}_{k^*}$. From this our claim follows immediately.

<u>Remark 1</u>. In Theorems 5.2.1 and 5.2.2 we discussed conditions for existence and uniqueness by allowing k in the pair (e,k) to be any element of $K(\tilde{D})$, and not just an element of some specified subset $K' \subset K(\tilde{D})$. This, however, is no loss of generality since the latter, more general situation is obtained, provided $\tilde{D}$ is suitably restricted. Indeed, if $K' \subset K(\tilde{D})$, $K' \neq \emptyset$, we can put $\tilde{\tilde{D}} = \bigcup_{x \in K'} [(x+N) \cap \tilde{D}]$; it is then easy to see that $K(\tilde{\tilde{D}}) = K'$.

In this context, note also that the theorems in Section 1.2 follow from the present ones on setting $\tilde{\tilde{D}} = N \cap D$.

The conditions appearing in Theorem 5.2.2 can be simplified if Z is a linear operator. Indeed, we have

<u>Theorem 5.2.3</u>. Let $\mathfrak{h} = [Z,N,M]$ be an abstract network over $\mathcal{L}$, let D be a linear subspace of $\mathcal{L}$ and let $Z : D \to \mathcal{L}$ be a linear operator. Moreover, let $D' \subset D$ be a linear subspace; let $d \in N \cap D$ be a fixed element and let $\tilde{D} = d+D'$. Then for each $(e,k) \in R(\tilde{D})$ there exists in $\tilde{D}$ a unique solution i of $\mathfrak{h}$ corresponding to (e,k), if and only if the operator PZ is 1-to-1 on $N \cap D'$. In this case,

$$i = i^* + (PZ)_o^{-1} P(e-Zi^*), \tag{5.2.7}$$

where $(PZ)_o^{-1}$ is the inverse of $PZ : N \cap D' \to (PZ)(N \cap D')$ and i^* is any chosen element in $\tilde{D}_k = (k+N) \cap (d+D')$.

Remark 2. The assumption that $\tilde{D}$ is a linear flat in $\mathcal{L}$ is motivated by the case of a network $\mathfrak{n}$ such that $\mathcal{L}$ is some function space, $\mathfrak{n}$ contains differentiators and some initial condition is prescribed for the current regime in $\mathfrak{n}$. In this context, see the Example in Section 3.2.

Proof of Theorem 5.2.3: Choose some $(e,k) \in R(\tilde{D})$; then $\tilde{D}_k \neq \phi$, and we can choose some $i^* \in \tilde{D}_k$. Observe also that, by our hypothesis, $d \in N$, $d \in D$, and $i^* \in D$, since $d+D' \subset d+D = D$. Moreover, we have

$$\tilde{D}_k = i^* + N \cap D'. \tag{5.2.8}$$

Indeed, let $x \in i^* + N \cap D'$; then $x = i^*+n$, where $n \in N$ and $n \in D'$. Since $i^* \in k+N$, we have $x \in n+k+N$, and because $n+N = N$, it follows that $x \in k+N$. Moreover, since $i^* \in d+D'$, we have $x \in n+d+D'$; however, $n+D' = D'$, so that $x \in d+D'$. Consequently, $x \in (k+N) \cap (d+D') = \tilde{D}_k$.

Conversely, let $x \in \tilde{D}_k$, i.e., $x \in k+N$ and $x \in d+D'$. Thus, there exists some $n \in N$ such that $x = k+n$, and some $q \in D'$ such that $x = d+q$. Furthermore, since $i^* \in k+N$ and $i^* \in d+D'$, there exists $n' \in N$ such that $i^* = k+n'$, and some $q' \in D'$ so that $i^* = d+q'$. Hence, $x-i^* = n-n' \in N$, and $x-i^* = q-q' \in D'$, i.e., $x-i^* \in N \cap D'$; consequently, $x \in i^*+N \cap D'$ and the relation (5.2.8) is proved.

Next, since $i^* \in D$ and $N \cap D' \subset D$, it follows from (5.2.8) by linearity of Z,

$$Q(\tilde{D}_k) = M+Z\tilde{D}_k = M+Zi^*+Z(N \cap D'). \tag{5.2.9}$$

Moreover, it is easy to see that the following statement is true:

PZ is 1-to-1 on $N \cap D'$ $\Leftrightarrow$ PZ is 1-to-1 on $\tilde{D}_k$ for any $k \in K(\tilde{D})$.

Indeed, let PZ be 1-to-1 on $N \cap D'$; assume that $PZx_1 = PZx_2$ for some $x_1, x_2 \in \tilde{D}_k$ for some chosen $k \in K(\tilde{D})$. Choose an $i^* \in \tilde{D}_k$; then, by (5.2.8), $x_j = i^*+y_j$ for some $y_j \in N \cap D'$, $j = 1,2$. Thus, we have $PZ(i^*+y_1) = PZ(i^*+y_2)$ which implies by linearity of Z that $PZy_1 = PZy_2$; hence, $y_1 = y_2$, and consequently, $x_1 = x_2$, i.e., PZ is 1-to-1 on $\tilde{D}_k$. The proof of the converse implication is similar.

However, this fact proves by Theorem 5.2.2 the first claim in our theorem.

Finally, since $(e,k) \in R(\tilde{D})$ and $i^* \in \tilde{D}_k$, it follows by (5.2.9) that $e \in Q(\tilde{D}_k) = M+Zi^* + Z(N \cap D')$; consequently, $P(e-Zi^*) \in P(M+Z(N \cap D')) = (PZ)(N \cap D')$. Thus, if PZ is 1-to-1 on $N \cap D'$, the element $(PZ)_o^{-1}P(e-Zi^*)$ is in $N \cap D'$, and consequently, by (5.2.8), $i = i^* + (PZ)_o^{-1}P(e-Zi^*)$ is in $\tilde{D}_k$, i.e., the element i defined by (5.2.7) satisfies (ii). Similarly, we have

$P(Zi-e) = PZ(i^*+(PZ)_o^{-1}P(e-Zi^*)) - Pe = 0$, i.e., (i) holds. Hence, i is the solution of $\mathfrak{h}$ corresponding to (e,k).

Let us now apply the previous results to Hilbert networks. First, let us introduce the following notation:

If H is a Hilbert space and $c \leq \aleph_o$, let H_c be the space of all c-vectors whose elements are in H.

Now, let $\hat{\mathfrak{h}} = (\hat{Z},G)$ be a Hilbert network with G being locally finite, and let d be the $c_2 \times c_1$ incidence matrix of G; also, let $\hat{a} \in [H^{c_2},H^{c_1}]$ be defined as in Section 2.1. Denote

$$d^T H^{c_2} = \{d^T \cdot x : x \in H^{c_2}\}. \tag{5.2.10}$$

Clearly, $d^T H^{c_2}$ is a linear subspace of H_{c_1}.

Let us make the following observation:

Let $k \in H^{c_2}$ and let $j = d^T \cdot k$; if $\xi \in H^{c_2}$, then $d^T \cdot \xi = j \Leftrightarrow \xi \in k+N_{\hat{a}}$.

Indeed, if $d^T \cdot \xi = j$, then $d^T \cdot (\xi - k) = 0$ so that $\hat{a}(\xi-k) = 0$ by (2.1.12); hence, $\xi - k \in N_{\hat{a}}$, i.e., $\xi \in k+N_{\hat{a}}$. Conversely, if $\xi \in k+N_{\hat{a}}$, then $\xi-k \in N_{\hat{a}}$, which implies by (2.1.12) that $d^T \cdot (\xi-k) = 0$; hence, $d^T \cdot \xi = d^T \cdot k = j$.

<u>Definition</u>. Let $\hat{\mathfrak{h}}$ be a Hilbert network and let $(e,j) \in H^{c_2} \times d^T H^{c_2}$; an element $i \in H^{c_2}$ will be called a solution of $\hat{\mathfrak{h}}$ corresponding to the pair (e,j), if i is a solution of the abstract network $\mathfrak{h} = [\hat{Z}, N_{\hat{a}}, N_{\hat{a}}^{\perp}]$ over H^{c_2} corresponding to $(e,k) \in H^{c_2} \times H^{c_2}$, where k is any element in H^{c_2} satisfying the equation $d^T \cdot k = j$.

In other words,

(i)' there exists $v \in \hat{Z}i$ such that $v-e \in N_{\hat{a}}^{\perp}$,

(ii)' $i \in (k+N_{\hat{a}}) \cap D$.

Observe that this definition is meaningful, i.e., i does not depend on the choice of k. Actually, if k' is another vector in H^{c_2} satisfying equation $d^T \cdot k' = j$, then, by the above observation, $k' \in k+N_{\hat{a}}$, i.e., $k' = k+n$, $n \in N_{\hat{a}}$; hence, $k'+N_{\hat{a}} = k+n+N_{\hat{a}} = k+N_{\hat{a}}$. Consequently, the linear flat $k+N_{\hat{a}}$ is completely determined by $j \in d^T H^{c_2}$.

<u>Definition</u>. Let $\hat{\mathfrak{h}}$ be a Hilbert network and let $(e,j) \in H^{c_2} \times d^T H^{c_2}$; an element $i \in H^{c_2}$ will be called a classical solution of $\hat{\mathfrak{h}}$ corresponding to the pair (e,j) if

K_1^+: there exists a $v \in \hat{Z}i$ such that $\bar{\gamma}^T \cdot (v-e) = 0$ for all $\gamma \in R^{c_2}$

satisfying the equation $d^T \cdot \gamma = 0$,

K_2^+: $i \in D$ and $d^T \cdot i = j$.

Then we have the same proposition as Theorem 2.2.1, i.e.,

Theorem 5.2.4. Let $(e,j) \in H^{c_2} \times d^T H^{c_2}$ and $i \in H^{c_2}$; then i is a solution of $\hat{n}$ corresponding to (e,j), if and only if i is a classical solution of $\hat{n}$ corresponding to (e,j).

The proof is practically the same as that of Theorem 2.2.1. Indeed, it is clear that $K_2^+ \Leftrightarrow$ (ii)'. Moreover, the equivalence $K_1^+ \Leftrightarrow$ (i)' follows by using the same argument as in the proof of Lemma 2.2.1.

Let us now prove a (rather complicated) analog of Theorem 2.2.2. To this end, let us introduce some new notation and carry out certain auxiliary considerations.

Let $\tilde{D} \subset D \subset H^{c_2}$, $\tilde{D} \neq \emptyset$; if $k \in H^{c_2}$, we will define the set $\tilde{D}_k$ again by (5.2.2), and $Q(\tilde{D}_k)$ by (5.2.3). (Here, of course, $M = N_{\hat{a}}^{\perp}$.) Also, let $K(\tilde{D})$ and $R(\tilde{D})$ be defined by (5.2.4) and (5.2.5), respectively. In addition to this, let

$$S(\tilde{D}) = \{(x, d^T \cdot y) : y \in K(\tilde{D}),\ x \in Q(\tilde{D}_y)\}. \tag{5.2.11}$$

If $\xi \in H^{c_2}$, let the operator $T_\xi : H^{c_2} \to H^{c_2}$ be defined by

$$T_\xi z = z + \xi. \tag{5.2.12}$$

Clearly, T_ξ is a homeomorphism.

Next, for every $x \in K(\tilde{D})$, (i.e., $\tilde{D}_x = (x+N_{\hat{a}}) \cap \tilde{D} \neq \emptyset$), choose a fixed $\xi_x \in \tilde{D}_x$ and define a subset $F_x \subset H^{c_o}$ by the relation

$$\hat{X} F_x = T_{-\xi_x} \tilde{D}_x. \tag{5.2.13}$$

This definition is clearly meaningful. Indeed, if $u \in T_{-\xi_x} \tilde{D}_x$, then $u = -\xi_x + v$, $v \in x+N_{\hat{a}}$, $v \in \tilde{D}$, so that $u \in -\xi_x + x + N_{\hat{a}}$. However, since $\xi_x \in x+N_{\hat{a}}$, i.e., $-\xi_x \in -x+N_{\hat{a}}$, we have $u \in -x+N_{\hat{a}}+x+N_{\hat{a}} = N_{\hat{a}}$. Hence, $T_{-\xi_x} \tilde{D}_x \subset N_{\hat{a}}$. On the other hand, by Lemma 2.1.2, $\hat{X}$ is a 1-to-1 correspondence between H^{c_o} and $N_{\hat{a}}$, and consequently, F_x is uniquely defined by (5.2.13).

Theorem 5.2.5. Let $\hat{n}$ be a H.n., let $\tilde{D} \subset D$, $\tilde{D} \neq \emptyset$, and let $(e,j) \in H^{c_2} \times d^T H^{c_2}$; then $\hat{n}$ possesses a solution $i \in \tilde{D}$ corresponding to (e,j), if and only if $(e,j) \in S(\tilde{D})$. In this case, the set I of all solutions in $\tilde{D}$ corresponding to (e,j) is given by

$$I = T_{\xi_k} \hat{X} W_{\xi_k}^{-} \hat{X}^{*\prime} e, \tag{5.2.14}$$

where k is any element in H^{c_2} satisfying the equation $d^T \cdot k = j$, $\xi_k \in \tilde{D}_k$, and where $W_{\xi_k}^{-}$ signifies the quasiinverse of $W_{\xi_k} : F_k \to \mathfrak{S}([W_{\xi_k} F_k]^{\circ})$ with W_{ξ_k} being defined by

$$W_{\xi_k} = \hat{X}^* \hat{Z} T_{\xi_k} \hat{X}. \tag{5.2.15}$$

<u>Proof</u>: (1) If $(e,j) \in S(\tilde{D})$, then $j = d^T \cdot y$ for some $y \in K(\tilde{D})$. By Theorem 5.2.1, there exists a solution $i \in \tilde{D}$ of the associated abstract network $\eta = (\hat{Z},\hat{a})$ corresponding to $(e,y) \in R(\tilde{D})$. However, by our definition, i is precisely a solution of $\hat{\eta}$ corresponding to (e,j).

(2) Conversely, let $i \in \tilde{D}$ be a solution of $\hat{\eta}$ corresponding to $(e,j) \in H^{c_2} \times d^T H^{c_2}$. Then, by our definition, i is a solution of the abstract network η corresponding to (e,k), where $k \in H^{c_2}$ and $d^T \cdot k = j$. By Theorem 5.2.1, $(e,k) \in R(\tilde{D})$, and consequently, $(e,j) = (e,d^T \cdot k) \in S(\tilde{D})$.

To prove formula (5.2.14), choose some $(e,j) \in S(\tilde{D})$, some $k \in H^{c_2}$ with $d^T \cdot k = j$, some $\xi_k \in \tilde{D}_k$ and define the set I' by

$$I' = T_{\xi_k} \hat{X} W_{\xi_k}^{-} \hat{X}^* e. \tag{5.2.16}$$

First, we are going to show that I' is well-defined. Indeed, since $(e,k) \in R(\tilde{D})$, we infer by (5.2.5) and (5.2.3) that there exist elements $m \in N_{\hat{a}}^{\perp} = N_{\hat{X}^*}$ and $n \in (\hat{Z}\tilde{D}_k)^{\circ}$ such that $e = m+n$. Because $\hat{X}^* m = \theta$, we have $\hat{X}^* e = \hat{X}^* n$, and consequently,

$$\hat{X}^* e \in \hat{X}^*(\hat{Z}\tilde{D}_k)^{\circ} = [\hat{X}^* \hat{Z} T_{\xi_k} (T_{-\xi_k} \tilde{D}_k)]^{\circ} = [\hat{X}^* \hat{Z} T_{\xi_k} \hat{X}(Y_k T_{-\xi_k} \tilde{D}_k)]^{\circ}$$
$$= [W_{\xi_k} F_k]^{\circ};$$

here, $Y_k : T_{-\xi_k} \tilde{D}_k \to F_k$ signifies the inverse of $\hat{X} : F_k \to T_{-\xi_k} \tilde{D}_k$. Hence, the element $\hat{X}^* e$ lies in the domain of $W_{\xi_k}^{-}$ so that the expression $T_{\xi_k} \hat{X} W_{\xi_k}^{-} \hat{X}^* e$ is meaningful.

Now, let $i \in I$. Then by (ii)', $i \in (k+N_{\hat{a}}) \cap D$, and since $i \in \tilde{D}$, we have $i \in \tilde{D}_k = T_{\xi_k} \hat{X} F_k$ by (5.2.13), i.e., there exists $u \in F_k$ such that $i = T_{\xi_k} \hat{X} u$. Moreover, by (i)', there exists $v \in \hat{Z} i$ such that $v - e \in N_{\hat{a}}^{\perp} = N_{\hat{X}^*}$; thus $\hat{X}^* e = \hat{X}^* v$. Hence, $\hat{X}^* v \in \hat{X}^* \hat{Z} i = \hat{X}^* \hat{Z} T_{\xi_k} \hat{X} u = W_{\xi_k} u$. This yields $u \in W_{\xi_k}^{-} \hat{X}^* v = W_{\xi_k}^{-} \hat{X}^* e$; consequently, $i = T_{\xi_k} \hat{X} u \in T_{\xi_k} \hat{X} W_{\xi_k}^{-} \hat{X}^* e = I'$. Hence, $I \subset I'$.

Conversely, let $i \in I'$. Then by (5.2.16) there exists $p \in W^-_{\xi_k} \hat{X}^* e$ such that $i = T_{\xi_k} \hat{X} p$. Since $W^-_{\xi_k} \hat{X}^* e \subset F_k$, it follows that $p \in F_k$, and consequently, $i \in T_{\xi_k} \hat{X} F_k = \tilde{D}_k$. Hence, i satisfies (ii)' and $i \in \tilde{D}$. On the other hand, by the above, $\hat{X}^* e \in W_{\xi_k} p = \hat{X}^* \hat{Z} T_{\xi_k} \hat{X} p = \hat{X}^* \hat{Z} i$; hence, there exists $v \in \hat{Z} i$ such that $\hat{X}^* e = \hat{X}^* v$. Thus, $v-e \in N_{\hat{X}^*} = N^{\perp}_{\hat{a}}$, i.e., i satisfies (i)'. Consequently, $i \in I$ so that $I' \subset I$; this completes the proof.

<u>Theorem 5.2.6</u>. Let $\hat{n}$ be a H.n. and let $\tilde{D} \subset D$, $\tilde{D} \neq \phi$; then for each $(e,j) \in S(\tilde{D})$ there exists in $\tilde{D}$ a unique solution of $\hat{n}$ corresponding to (e,j), if and only if for every $x \in K(\tilde{D})$ and some $\xi_x \in \tilde{D}_x$ the set mapping $W_{\xi_x} = \hat{X}^* \hat{Z} T_{\xi_x} \hat{X}$ is simple on F_x.

<u>Proof</u>: Choose $x \in K(\tilde{D})$ and $\xi_x \in \tilde{D}_x$. Let $Y_x : T_{-\xi_x} \tilde{D}_x \to F_x$ be the inverse of $\hat{X} : F_x \to T_{-\xi_x} \tilde{D}_x$. Then we have by Lemma 1.1.3: W_{ξ_x} is simple on $F_x \Leftrightarrow \hat{X}^* \hat{Z} T_{\xi_x} = W_{\xi_x} Y_x$ is simple on $T_{-\xi_x} \tilde{D}_x \Leftrightarrow \hat{X}^* \hat{Z} = (\hat{X}^* \hat{Z} T_{\xi_x}) T_{-\xi_x}$ is simple on $\tilde{D}_x$ $\Leftrightarrow P\hat{Z} = \hat{X}\hat{X}^* \hat{Z}$ is simple on $\tilde{D}_x$. This equivalence and Theorem 5.2.2 concludes the proof.

<u>Theorem 5.2.7</u>. Let $\hat{n}$ be a H.n., let D be a linear subspace of H^{c_2} and let $\hat{Z} : D \to H^{c_2}$ be a linear operator. Moreover, let $D' \subset D$ be a linear subspace, let $d \in N_{\hat{a}} \cap D$ be a fixed element and let $\tilde{D} = d + D'$. Then for each $(e,j) \in S(\tilde{D})$ there exists in $\tilde{D}$ a unique solution i of $\hat{n}$ corresponding to (e,j), if and only if the operator $W = \hat{X}^* \hat{Z} \hat{X}$ is 1-to-1 on F', where the linear subspace $F' \subset H^{c_o}$ is defined by

$$\hat{X} F' = N_{\hat{a}} \cap D'. \tag{5.2.17}$$

In this case,

$$i = i^* + \hat{X} W^{-1} \hat{X}^* (e - \hat{Z} i^*), \tag{5.2.18}$$

where W^{-1} denotes the inverse of $W : F' \to WF'$, i^* is any chosen element in $\tilde{D}_k = (k + N_{\hat{a}}) \cap (d + D')$ and $k \in H^{c_2}$ is any solution of $d^T \cdot k = j$.

<u>Proof</u>: The first statement of the theorem follows immediately from Theorem 5.2.3 and from the equivalence: $\hat{X}^* \hat{Z} \hat{X}$ is 1-to-1 on $F' \Leftrightarrow P\hat{Z}$ is 1-to-1 on $N_{\hat{a}} \cap D'$.

To prove (5.2.18) it suffices to realize that, under the assumptions made, $e - \hat{Z} i^* \in N^{\perp}_{\hat{a}} + Z(N_{\hat{a}} \cap D') = P^{-1}\{P\hat{Z}(N_{\hat{a}} \cap D')\}$, and that $(P\hat{Z})^{-1} P = \hat{X} W^{-1} \hat{X}^*$ on $P^{-1}\{P\hat{Z}(N_{\hat{a}} \cap D')\}$; the rest follows from Theorem 5.2.3.

Remark 3. The Theorem 5.2.7 clearly generalizes the rule of classical network theory stating that current sources in a linear network can be replaced by additional voltage sources, [6].

Note also that sufficient conditions for uniqueness of a solution can easily be derived on the basis of conditions stated in Theorem 1.2.5. For example, if any one of conditions (1.2.5), (1.2.7), (1.2.10) is satisfied for all $x_1, x_2 \in H^{c_2}$, then we have uniqueness of solutions in D for any $(e,j) \in S(D)$; this follows immediately from Theorem 5.2.2 and the fact that $\tilde{D}_k \subset D$ for any $k \in K(D)$. Since these conclusions are straightforward, we omit the details.

CHAPTER 6

FINITE NETWORKS WITHOUT TOPOLOGY

In many problems concerning specific networks the network variables (voltages, currents) are, for example, continuous functions on $[0,\infty)$, or locally integrable functions, etc. Thus, the network variables cannot be embedded in any Hilbert space, and we are unable to use the model of the Hilbert network developed in Chapter 2. Fortunately, several results obtained in Chapter 2 can be generalized to the case when we assume only that the network variables belong to some linear space not equipped with any topology. However, for this generalization we have to pay a price: it has to be assumed that the networks in question are finite. Also, because we will not make use of the Hilbert space structure, we will not be able to prove such simple yet powerful theorems as Theorem 2.2.5 or 2.2.6. In other words, our theorems will have a purely algebraic character, and consequently, their practical applicability will be limited to rather simple situations of networks, such as networks with constant elements, etc.

6.1 Algebraic Networks

In this section, we will consider finite networks whose underlying space is a linear space without any topology.

First, let us establish some preliminary results concerning finite graphs. We will use the same notation as in Section 2.1 although the meaning of some symbols will be different; we believe that this approach will not lead to any misunderstanding, and, at the same time, will facilitate memorizing the various concepts.

Let G be a finite oriented graph (see Section 2.1), i.e., $c_1, c_2 < \aleph_0$, and let the $c_2 \times c_1$ matrix d be defined by (2.1.3). Thus, let us emphasize that d is a finite matrix.

Define the mappings $\tilde{a}$ and $\tilde{a}^*$ by

$$\tilde{a} : R^{c_2} \to R^{c_1}, \qquad \tilde{a}\xi = d^T \cdot \xi, \tag{6.1.1}$$

$$\tilde{a}^* : R^{c_1} \to R^{c_2}, \qquad \tilde{a}^*\eta = d\cdot\eta,$$

and let $N_{\tilde{a}} = \{\xi : \xi \in R^{c_2}, \tilde{a}\xi = 0\}$.

As in Section 2.1 we will assume that $N_{\tilde{a}} \neq \{0\}$, which amounts to requiring that G contains at least one loop.

In the present considerations we will take advantage of the fact that every space R^c, $c < \aleph_0$ is a Hilbert space with inner product $\langle \xi_1, \xi_2 \rangle = \xi_1^T \cdot \bar{\xi}_2$. Thus, since $N_{\tilde{a}}$ is a closed subspace of R^{c_2}, let us choose a fixed orthonormal basis $\{\xi^1, \xi^2, \ldots, \xi^{c_0}\}$ in $N_{\tilde{a}}$, and let X be the $c_2 \times c_0$ matrix having the c_2-vectors ξ^i as columns.

Furthermore, define mappings $\tilde{X}$ and $\tilde{X}^*$ by

$$\tilde{X} : R^{c_0} \to R^{c_2}, \qquad \tilde{X}\xi = X\cdot\xi, \tag{6.1.2}$$

$$\tilde{X}^* : R^{c_2} \to R^{c_0}, \qquad \tilde{X}^*\eta = \bar{X}^T\cdot\eta,$$

and let $N_{\tilde{X}^*} = \{\eta : \eta \in R^{c_2}, \tilde{X}^*\eta = 0\}$.

Then we have

<u>Lemma 6.1.1.</u>

(i) $N_{\tilde{a}} = \tilde{X}R^{c_0}$ and $\tilde{X}$ is 1-to-1,

(ii) $\tilde{a}\tilde{X} = 0$ on R^{c_0},

(iii) $\tilde{X}^*\tilde{X} = I$ on R^{c_0},

(iv) $N_{\tilde{X}^*} = \tilde{a}^*R^{c_1}$,

(v) $R^{c_2} = N_{\tilde{X}^*} \oplus N_{\tilde{a}}$.

<u>Proof</u>: Propositions (i), (ii), (iii) are trivial consequences of the definitions. Realizing the fact that R^{c_0}, R^{c_1}, R^{c_2} are Hilbert spaces, and that, by our definition, $\tilde{a}^*$ is the adjoint of $\tilde{a}$, and $\tilde{X}^*$ is the adjoint of $\tilde{X}$, we have by Lemma 1.4.2, and (i),

$$N_{\tilde{X}^*} = (\tilde{X}R^{c_0})^\perp = N_{\tilde{a}}^\perp, \tag{6.1.3}$$

and

$$N_{\tilde{a}} = (\tilde{a}^*R^{c_1})^\perp. \tag{6.1.4}$$

Hence, by Lemma 1.4.1, $N_{\tilde{X}^*} = (\tilde{a}^*R^{c_1})^{\perp\perp} = \overline{(\tilde{a}^*R^{c_1})}$. However, $\tilde{a}^*R^{c_1}$ is a <u>finite</u> dimensional subspace of R^{c_2}, and consequently, it is closed; thus $\overline{(\tilde{a}^*R^{c_1})} = \tilde{a}^*R^{c_1}$ and (iv) is proved.

Furthermore, since $R^{c_2} = N_{\tilde{a}}^{\perp} \oplus N_{\tilde{a}}$, (6.1.3) proves (v).

Next, let L be a fixed nontrivial linear space; if $c \geq 1$ is an integer, L^c will denote the Cartesian product $L \times L \times \ldots \times L$ having c factors. Clearly, L^c is a real or complex linear space with "elementwise" operations, depending on whether L is real or complex. For convenience, we will interpret the elements in L^c as c-vectors.

Define now the mappings $\hat{a}$ and $\hat{a}^*$ by

$$\hat{a} : L^{c_2} \to L^{c_1}, \qquad \hat{a}x = d^T \cdot x, \tag{6.1.5}$$

$$\hat{a}^* : L^{c_1} \to L^{c_2}, \qquad \hat{a}^*y = d \cdot y,$$

and let $N_{\hat{a}} = \{x : x \in L^{c_2}, \hat{a}x = 0\}$.

Moreover, define the mappings $\hat{X}$ and $\hat{X}^*$ by

$$\hat{X} : L^{c_0} \to L^{c_2}, \qquad \hat{X}x = X \cdot x, \tag{6.1.6}$$

$$\hat{X}^* : L^{c_2} \to L^{c_0}, \qquad \hat{X}^*y = \bar{X}^T \cdot y,$$

and let $N_{\hat{X}^*} = \{y : y \in L^{c_2}, \hat{X}^*y = 0\}$.

In definitions (6.1.6) it is understood that if L is a real space, then spaces $R^{c_0}, R^{c_1}, R^{c_2}$ are assumed to be real Euclidean spaces, and consequently, X is a real matrix; if, on the other hand, L is complex, X can be a complex matrix.

Using Lemma 6.1.1, we can easily prove the following result:

Lemma 6.1.2.

(i) $N_{\hat{a}} = \hat{X}L^{c_0}$ and $\hat{X}$ is 1-to-1;

(ii) $\hat{a}\hat{X} = 0$ on L^{c_0};

(iii) $\hat{X}^*\hat{X} = I$ on L^{c_0};

(iv) $N_{\hat{X}^*} = \hat{a}^*L^{c_1}$;

(v) $L^{c_2} = N_{\hat{X}^*} \oplus N_{\hat{a}}$;

(vi) if P is the projection from L^{c_2} onto $N_{\hat{a}}$ along $N_{\hat{X}^*}$, then $P = \hat{X}\hat{X}^*$;

(vii) let $u \in L^{c_2}$; then $\bar{\gamma}^T \cdot u = 0$ for all $\gamma \in N_{\tilde{a}} \Leftrightarrow u \in N_{\hat{X}^*}$.

Proof: First of all let us emphasize the fact that all the operators defined above are defined via finite matrices with constant elements; consequently, the product of such matrices obeys the associative law. Having this fact in mind, assertions (ii) and (iii) are trivial consequences of the definitions.

To prove (i), choose some $x \in N_{\hat{a}}$; then $\hat{a}x = d^T \cdot x = 0$. Let $\mathfrak{B}$ be some basis in L. Since x has finitely many components, there exists finitely many elements $\ell_1, \ell_2, \ldots, \ell_m \in \mathfrak{B}$ and a $c_2 \times m$ matrix R with constant elements such that $x = R \cdot \ell$, where the m-vector ℓ is defined by $\ell^T = [\ell_1, \ell_2, \ldots, \ell_m]$. Thus, we have

$$\hat{a}x = d^T \cdot (R \cdot \ell) = (d^T \cdot R) \cdot \ell = 0. \tag{6.1.7}$$

Since the elements $\ell_1, \ell_2, \ldots, \ell_m$ are linearly independent, it follows from (6.1.7) that necessarily

$$d^T \cdot R = 0. \tag{6.1.8}$$

Hence, denoting the columns of R by $r^1, r^2, \ldots, r^m$, we have $d^T \cdot r^k = 0$ for $k = 1,2,\ldots,m$. However, by Lemma 6.1.1 (i) and (6.1.1), there exists $\xi^k \in R^{c_o}$, $k = 1,2,\ldots,m$, such that $r^k = X \cdot \xi^k$; thus, letting E be the $c_o \times m$ matrix whose columns are the ξ^k's, we get $R = X \cdot E$. Consequently, $x = (X \cdot E) \cdot \ell = X \cdot (E \cdot \ell)$, and since $E \cdot \ell \in L^{c_o}$, we get by (6.1.6), $x \in \hat{X}L^{c_o}$; hence, $N_{\hat{a}} \subset \hat{X}L^{c_o}$.

Conversely, let $x \in \hat{X}L^{c_o}$; then $x = \hat{X}y$ for some $y \in L^{c_o}$, and thus, by (ii), $\hat{a}x = \hat{a}\hat{X}y = 0$, i.e., $x \in N_{\hat{a}}$. Hence, $\hat{X}L^{c_o} \subset N_{\hat{a}}$.

To prove that $\hat{X}$ is 1-to-1, suppose that $\hat{X}x = 0$ for some $x \in L^{c_o}$. As above, there exists $\ell_1, \ldots, \ell_n \in \mathfrak{B}$ and a $c_o \times n$ matrix R' such that $x = R' \cdot \ell'$ with $\ell'^T = [\ell_1, \ell_2, \ldots, \ell_n]$. Thus, $\hat{X}x = X \cdot (R' \cdot \ell') = (X \cdot R') \cdot \ell = 0$ which yields, due to linear independence of the ℓ_i's, that $X \cdot R' = 0$. However, Lemma 6.1.1 (i) shows that each column of R' must be zero, i.e., $R' = 0$; hence, $x = 0$, which proves that $\hat{X}$ is 1-to-1.

To prove (iv), assume that $x \in N_{\hat{X}*}$, i.e., $\bar{X}^T \cdot x = 0$. For the vector $x \in L^{c_2}$, there exists $\ell_1, \ell_2, \ldots, \ell_m \in \mathfrak{B}$ and a matrix Q of type $c_2 \times m$ such that $x = Q \cdot \ell$ with $\ell^T = [\ell_1, \ell_2, \ldots, \ell_m]$. Thus, we have $\bar{X}^T \cdot (Q \cdot \ell) = (\bar{X}^T \cdot Q) \cdot \ell = 0$, which implies that $\bar{X}^T \cdot Q = 0$. However, by Lemma 6.1.1 (iv), each column q^i of Q can be written as $q^i = d \cdot \eta^i$, where $\eta^i \in R^{c_1}$; thus, defining the $c_1 \times m$ matrix E' by $E' = [\eta^1, \eta^2, \ldots, \eta^m]$, we have $Q = d \cdot E'$. Hence, $x = Q \cdot \ell = (d \cdot E') \cdot \ell = d \cdot (E' \cdot \ell)$, and because $E' \cdot \ell \in L^{c_1}$, we have $x \in \hat{a}*L^{c_1}$; thus, $N_{\hat{X}*} \subset \hat{a}*L^{c_1}$.

Conversely, let $x \in \hat{a}*L^{c_1}$, i.e., $x = \hat{a}*y$ for some $y \in L^{c_1}$; then we have $\hat{X}*x = \bar{X}^T \cdot (d \cdot y) = (\bar{X}^T \cdot d) \cdot y$. However, by definition of matrix X we have $d^T \cdot X = 0$, and since d is real, $d^T \cdot \bar{X} = 0$. Hence, $\bar{X}^T \cdot d = 0$ so that $\hat{X}*x = 0$, i.e., $x \in N_{\hat{X}*}$. Consequently, $\hat{a}*L^{c_1} \subset N_{\hat{X}*}$, which proves (iv).

As for (v), let $x \in L^{c_2}$. Then there exists $\ell_1, \ell_2, \ldots, \ell_p \in \mathfrak{B}$ and vectors $\xi_1, \xi_2, \ldots, \xi_p \in R^{c_2}$ such that

$$x = \sum_{i=1}^{p} (\xi_i \ell_i). \tag{6.1.9}$$

However, by Lemma 6.1.1 (v), for each i there exists $\xi_i' \in N_{\tilde{X}*}$ and $\xi_i'' \in N_{\tilde{a}}$ such that $\xi_i = \xi_i' + \xi_i''$. Put

$$x' = \sum_{i=1}^{p} (\xi_i' \ell_i), \quad x'' = \sum_{i=1}^{p} (\xi_i'' \ell_i); \tag{6.1.10}$$

then we have $x = x' + x''$, and also

$$\hat{X}*\xi_i' = \bar{X}^T \cdot \xi_i' = 0, \quad \tilde{a}\xi_i'' = d^T \cdot \xi_i'' = 0 \tag{6.1.11}$$

for every i. Thus, by (6.1.10) and definition of $\hat{X}*$ and $\hat{a}$,

$$\hat{X}*x' = \bar{X}^T \cdot x' = \sum_{i=1}^{p} (\bar{X}^T \cdot \xi_i')\ell_i = 0,$$

and

$$\hat{a}x'' = d^T \cdot x'' = \sum_{i=1}^{p} (d^T \cdot \xi_i'')\ell_i = 0,$$

i.e., $x' \in N_{\hat{X}*}$ and $x'' \in N_{\hat{a}}$. So, we proved that $L^{c_2} = N_{\hat{X}*} + N_{\hat{a}}$. To prove that this is actually a direct sum, assume that $x \in N_{\hat{X}*} \cap N_{\hat{a}}$. Then $\hat{a}x = 0$ and $\hat{X}*x = 0$. However, by (i) there exists $y \in L^{c_o}$ such that $x = \hat{X}y$, and consequently, by (iii), $0 = \hat{X}*x = \hat{X}*\hat{X}y = y$; thus, $x = 0$ so that $N_{\hat{X}*} \cap N_{\hat{a}} = \{0\}$. Hence, $L^{c_2} = N_{\hat{X}*} \oplus N_{\hat{a}}$ which concludes the proof of (v).

In order to prove (vi), note first that $\hat{X}*$ maps L^{c_2} onto L^{c_o}, because, according to (iii), $\hat{X}$ is the right-inverse of $\hat{X}*$. Next, define $Q : L^{c_2} \to L^{c_2}$ by $Q = \hat{X}\hat{X}*$. Then we have by (iii), $Q^2 = \hat{X}\hat{X}*\hat{X}\hat{X}* = \hat{X}\hat{X}* = Q$, i.e., Q is a projection.

Moreover, let $x \in QL^{c_2}$. Then $x = \hat{X}\hat{X}*y$ for some $y \in L^{c_2}$, and we have by (ii), $\hat{a}x = (\hat{a}\hat{X})\hat{X}*y = 0$. Hence, $QL^{c_2} \subset N_{\hat{a}}$. Conversely, let $x \in N_{\hat{a}}$. Then, by (i), $x = \hat{X}z$ for some $z \in L^{c_o}$, and since $\hat{X}*$ maps L^{c_2} onto L^{c_o}, there exists $w \in L^{c_2}$ such that $z = \hat{X}*w$. Thus, $x = \hat{X}\hat{X}*w = Qw$, i.e., $N_{\hat{a}} \subset QL^{c_2}$. Hence, $QL^{c_2} = N_{\hat{a}}$ which shows by (v) that Q is the projection onto $N_{\hat{a}}$ along $N_{\hat{X}*}$; thus, (vi) is proven.

Finally, let $u \in L^{c_2}$ and assume that $\bar{\gamma}^T \cdot u = 0$ for all $\gamma \in N_{\tilde{a}}$. Since each column ξ^i of X is in $N_{\tilde{a}}$, we have $\bar{\xi}^{i^T} \cdot u = 0$; consequently, $\bar{X}^T \cdot u = 0$, i.e., $u \in N_{\hat{X}*}$.

Conversely, let $u \in N_{\hat{X}*}$. Then $\bar{X}^T \cdot u = 0$. Now, if $\gamma \in N_{\tilde{a}}$, then, by Lemma 6.1.1, there exists $\eta \in R^{c_o}$ such that $\gamma = X \cdot \eta$; hence, we have $0 = \bar{\eta}^T \cdot (\bar{X}^T \cdot u) = (\bar{\eta}^T \cdot \bar{X}^T) \cdot u = (\overline{X \cdot \eta})^T \cdot u = \bar{\gamma}^T \cdot u$. This concludes the proof of

(vii) and of our Lemma.

Now we are ready to introduce the algebraic networks. Let G be a finite oriented graph, and let the symbols $\hat{a}$, $\hat{a}*$, $\hat{X}$, $\hat{X}*$ etc. have the meaning defined above. Moreover, let $D \subset L^{c_2}$, $D \neq \emptyset$, and let $\hat{Z} : D \to \mathfrak{S}(L^{c_2})$ be a set mapping. Then the ordered pair $\hat{n} = (\hat{Z},G)$ will be called an algebraic network.

Definition. Let $\hat{n} = (\hat{Z},G)$ be an algebraic network, and let $e \in L^{c_2}$; an element $i \in L^{c_2}$ will be called a solution of $\hat{n}$ corresponding to e, if

$\tilde{K}_1$: there exists an element $v \in \hat{Z}i$ such that

$$\bar{\gamma}^T \cdot (v-e) = 0 \qquad (6.1.12)$$

for every $\gamma \in R^{c_2}$ satisfying the equation $d^T \cdot \gamma = 0$,

$\tilde{K}_2$: $i \in D$ and $d^T \cdot i = 0$.

From this definition we see readily that the following statement is true.

Theorem 6.1.1. Let $\hat{n} = (\hat{Z},G)$ be an algebraic network, let $e \in L^{c_2}$ and assume that $N_{\tilde{a}} \neq \{0\}$; then an element $i \in L^{c_2}$ is a solution of $\hat{n}$ corresponding to e, if and only if i is a solution of the abstract network $n = [\hat{Z}, N_{\hat{a}}, N_{\hat{X}*}]$ over L^{c_2} corresponding to e.

Proof: First of all, n is truly an abstract network over L^{c_2}, since $L^{c_2} = N_{\hat{a}} \oplus N_{\hat{X}*}$ by Lemma 6.1.2 (v). Moreover, it is clear that $\tilde{K}_2 \Leftrightarrow i \in D$ and $i \in N_{\hat{a}} \Leftrightarrow i \in N_{\hat{a}} \cap D \Leftrightarrow$ (ii) in the definition of a solution of an a.n. Also, denoting $u = v-e$, we have by Lemma 6.1.2 (vii), $\tilde{K}_1 \Leftrightarrow \bar{\gamma}^T \cdot u = 0$ for every $\gamma \in N_{\tilde{a}} \Leftrightarrow u \in N_{\hat{X}*} \Leftrightarrow$ condition (i) for an a.n. Hence, the proof.

Due to the above theorem, we can now easily state conditions for the existence and uniqueness of a solution of an algebraic network. Also, since $\hat{n}$ and n are formally identical concepts, we will understand by "regularity on D" or "admittance A" of $\hat{n}$, the corresponding concept for n defined in Section 1.2.

As in Section 2.2, it will be convenient to introduce the following set:

If $D \subset L^{c_2}$, $D \neq \emptyset$, let $F \subset L^{c_0}$ be defined by

$$\hat{X}F = N_{\hat{a}} \cap D. \tag{6.1.13}$$

This definition is meaningful due to (i) in Lemma 6.1.2. We will put $F = \emptyset \Leftrightarrow N_{\hat{a}} \cap D = \emptyset$. By Lemma 6.1.2 (iii), $F = \hat{X}^*(N_{\hat{a}} \cap D)$.

Theorem 6.1.2. Let $\hat{n} = (\hat{Z}, G)$ be an algebraic network, let $F \neq \emptyset$ and let $e \in L^{c_2}$; then $\hat{n}$ possesses a solution i corresponding to e, if and only if $e \in Q(D)$, where

$$Q(D) = N_{\hat{X}^*} + (\hat{Z}\hat{X}F)^{\circ}. \tag{6.1.14}$$

In this case, the set I of all solutions corresponding to e is given by $I = Ae$, where $A : Q(D) \to \mathfrak{S}(\hat{X}F)$ is the admittance of $\hat{n}$. Moreover,

$$A = \hat{X}W^{-}\hat{X}^*, \tag{6.1.15}$$

where W^{-} denotes the quasiinverse of the set mapping $W = \hat{X}^*\hat{Z}\hat{X} : F \to \mathfrak{S}((WF)^{\circ})$.

Proof: The first assertion is a trivial consequence of Theorem 6.1.1, 1.2.1 and (6.1.13). As for the second assertion, choose arbitrarily $e \in Q(D)$ and denote

$$I' = Ae = \hat{X}W^{-}\hat{X}^*e. \tag{6.1.16}$$

We can easilyverify that the set I' is well defined. Indeed, if $e \in Q(D)$, then, by (6.1.14), there exists $m \in N_{\hat{X}^*}$ and $n \in (\hat{Z}\hat{X}F)^{\circ}$ such that $e = m+n$. Because $\hat{X}^*m = 0$, we have

$$\hat{X}^*e = \hat{X}^*n \in \hat{X}^*(\hat{Z}\hat{X}F)^{\circ} = (WF)^{\circ}; \tag{6.1.17}$$

hence, $\hat{X}^*e$ is in the domain of W^{-}, and consequently, the product $\hat{X}W^{-}\hat{X}^*e$ is meaningful.

Now, let $i \in I$. Then by $\tilde{K}_2$, $i \in N_{\hat{a}} \cap D = \hat{X}F$, i.e., there exists $u \in F$ such that $i = \hat{X}u$. Furthermore, by $\tilde{K}_1$, (see the proof of Theorem 6.1.1), there exists $v \in \hat{Z}i$ such that $v-e \in N_{\hat{X}^*}$; thus

$$\hat{X}^*e = \hat{X}^*v. \tag{6.1.18}$$

Also, $\hat{X}^*v \in \hat{X}^*\hat{Z}i = \hat{X}^*\hat{Z}\hat{X}u = Wu$; this yields $u \in W^{-}\hat{X}^*v = W^{-}\hat{X}^*e$, so that $i = \hat{X}u \in \hat{X}W^{-}\hat{X}^*e = I'$. Hence, $I \subset I'$.

Conversely, let $i \in I'$. Then by (6.1.16), there exists $p \in W^{-}\hat{X}^*e$ such that $i = \hat{X}p$. Since $W^{-}\hat{X}^*e \subset F$, we have $p \in F$, and consequently, $i \in \hat{X}F = N_{\hat{a}} \cap D$, i.e., the element i satisfies $\tilde{K}_2$. On the other hand, the above relation yields $\hat{X}^*e \in Wp = \hat{X}^*\hat{Z}\hat{X}p = \hat{X}^*\hat{Z}i$; hence, there exists $v \in \hat{Z}i$ such that (6.1.18) holds. Then, $v-e \in N_{\hat{X}^*}$ and $\tilde{K}_1$ holds. Hence, $i \in I$ so that $I' \subset I$ and the proof is complete.

Theorem 6.1.3. Let $\hat{n} = (\hat{Z},G)$ be an algebraic network, and let $F \neq \phi$; then $\hat{n}$ is regular on D, if and only if the set mapping $W = \hat{X}^*\hat{Z}\hat{X}$ is simple on F.

Proof: Referring to Lemma 6.1.2, let $Y : N_{\hat{a}} \cap D \to F$ be the inverse of $\hat{X} : F \to N_{\hat{a}} \cap D$. Noting that $P = \hat{X}\hat{X}^*$, we have by Lemma 1.1.3: W is simple on $F \Leftrightarrow \hat{X}^*\hat{Z} = WY$ is simple on $N_{\hat{a}} \cap D \Leftrightarrow P\hat{Z} = \hat{X}\hat{X}^*\hat{Z}$ is simple on $N_{\hat{a}} \cap D$. This and Theorem 1.2.2 conclude the proof.

Theorem 6.1.4. Let $\hat{n} = (\hat{Z},G)$ be an algebraic network, and let $F \neq \phi$; then

$$Q(D) = L^{c_2} \Leftrightarrow (WF)^o = L^{c_o}, \tag{6.1.19}$$

where $W = \hat{X}^*\hat{Z}\hat{X}$.

Corollary. If $D \supset N_{\hat{a}}$, then $Q(D) = L^{c_2}$, if and only if $(WL^{c_o})^o = L^{c_o}$.

Proof: Using Theorem 1.2.3 and Lemma 6.1.2, we have: $Q(D) = L^{c_2} \Leftrightarrow [(P\hat{Z})(N_{\hat{a}} \cap D)]^o = N_{\hat{a}} \Leftrightarrow [\hat{X}\hat{X}^*\hat{Z}\hat{X}F]^o = \hat{X}L^{c_o} \Leftrightarrow \hat{X}(WF)^o = \hat{X}L^{c_o} \Leftrightarrow (WF)^o = L^{c_o}$. The corollary is obvious.

Remark 1. We can easily verify that, in formulas (6.1.14), (6.1.15) and (6.1.19), the operator $\hat{X}$ can be replaced by an operator generated by a matrix whose columns constitute a (not necessarily orthonormal) basis in $N_{\tilde{a}}$. (See also Problem 2.1.1 and 2.2.1.)

Indeed, let Y be any fixed $c_2 \times c_o$ matrix whose columns form a basis in $N_{\tilde{a}}$. Then there exists a nonsingular $c_o \times c_o$ matrix S such that $X = Y \cdot S$. Define operators $\hat{Y} : L^{c_o} \to L^{c_2}$, $\hat{Y}^* : L^{c_2} \to L^{c_o}$, $\hat{S} : L^{c_o} \to L^{c_o}$, $\hat{S}^* : L^{c_o} \to L^{c_o}$ by $\hat{Y}x = Y \cdot x$, $\hat{Y}^*y = \bar{Y}^T \cdot y$, $\hat{S}z = S \cdot z$, $\hat{S}^*u = \bar{S}^T \cdot u$. Then we clearly have $\hat{X} = \hat{Y}\hat{S}$, and consequently, $\hat{Y}$ is 1-to-1; also, $\hat{X}^* = \hat{S}^*\hat{Y}^*$.

Now, $N_{\hat{X}^*} = N_{\hat{Y}^*}$, because: $x \in N_{\hat{X}^*} \Leftrightarrow \bar{X}^T \cdot x = 0 \Leftrightarrow ((\bar{S}^T)^{-1} \cdot \bar{X}^T) \cdot x = 0 \Leftrightarrow \bar{Y}^T \cdot x = 0 \Leftrightarrow \hat{Y}^*x = 0 \Leftrightarrow x \in N_{\hat{Y}^*}$. Moreover, defining $F' \subset L^{c_o}$ by the relation

$$\hat{Y}F' = N_{\hat{a}} \cap D,$$

it follows by (6.1.13) that $\hat{X}F = \hat{Y}F'$. Hence, we have by (6.1.14),

$$Q(D) = N_{\hat{Y}^*} + (\hat{Z}\hat{Y}F')^o. \tag{6.1.20}$$

On the other hand, from Lemma 1.1.3 it follows easily that $\hat{X}*\hat{Z}\hat{X}$ is simple on F $\Leftrightarrow$ $\hat{Y}*\hat{Z}\hat{Y}$ is simple on F'. Also, $(\hat{X}*\hat{Z}\hat{X})^- = \hat{S}^{-1}(\hat{Y}*\hat{Z}\hat{Y})^-\hat{S}*^{-1}$. Hence, $\hat{X}(\hat{X}*\hat{Z}\hat{X})^-\hat{X}* = \hat{Y}\hat{S}\hat{S}^{-1}(\hat{Y}*\hat{Z}\hat{Y})^-\hat{S}*^{-1}\hat{S}*\hat{Y} = \hat{Y}(\hat{Y}*\hat{Z}\hat{Y})^-\hat{Y}*$. Thus, by (6.1.15),

$$A = \hat{Y}(\hat{Y}*\hat{Z}\hat{Y})^-\hat{Y}* \tag{6.1.21}$$

and our claim is confirmed.

Let us now establish a generalization of the classical Tellegen's theorem [6]. To this end, we will need some further concepts.

Let L' be a fixed linear space over the same systems of scalars as L has (i.e., either complex or real numbers), and let (L,L') signify the linear space of all linear operators from L into L'; moreover, let $\mathcal{K}$ be a fixed linear subspace of (L,L').

The subspace $\mathcal{K}$ will be called complete, if $x \in L$, $Ax = 0$ for all $A \in \mathcal{K}$ implies that $x = 0$.

Observe that if $\mathcal{K}$ contains at least one 1-to-1 operator, then $\mathcal{K}$ is complete.

If d is the matrix defined by (2.1.3) (the incidence matrix of G), let

$$\mathfrak{N} = \{\Gamma : \Gamma \in \mathcal{K}^{c_2},\ d^T\cdot\Gamma = 0\}. \tag{6.1.22}$$

Moreover, let the $c_2 \times c_o$ matrix X have the usual meaning; then we have

<u>Lemma 6.1.3</u>. $\Gamma \in \mathfrak{N} \Leftrightarrow \Gamma = X\cdot\omega$, where $\omega \in \mathcal{K}^{c_o}$.

<u>Proof</u>: Recalling the definition (6.1.5) of $\hat{a}$, and definition (6.1.6) of $\hat{X}$, we see that our proposition is nothing else than assertion (i) in Lemma 6.1.2, where the linear space L is replaced by $\mathcal{K}$.

<u>Theorem 6.1.5</u>. Let $\hat{n} = (\hat{Z},G)$ be an algebraic network, and let $e \in L^{c_2}$.

(i) Let $i \in L^{c_2}$ be a solution of $\hat{n}$ corresponding to e; then there exists $v \in \hat{Z}i$ such that

$$\Gamma^T\cdot(v-e) = 0 \tag{6.1.23}$$

for every $\Gamma \in \mathcal{K}^{c_2}$ satisfying the equation $d^T\cdot\Gamma = 0$.

(ii) Let $\mathcal{K}$ be complete, let $i \in D$ and $d^T\cdot i = 0$; if there exists $v \in \hat{Z}i$ such that (6.1.23) holds for every $\Gamma \in \mathcal{K}^{c_2}$ satisfying the equation $d^T\cdot\Gamma = 0$, then i is a solution of $\hat{n}$ corresponding to e.

<u>Proof</u>: First of all, observe the following essential fact: Since d is a

real matrix, we can clearly select an orthonormal basis $\{\xi^1, \xi^2, \ldots, \xi^{c_o}\}$ in $N_{\tilde{a}}$ so that each vector ξ^i is real, independently of whether R^{c_2} is a real or complex space; consequently, we can assume that X is a real matrix.

(i) If $i \in L^{c_2}$ is a solution of $\hat{n}$ corresponding to e, then, by $\tilde{K}_1$, there exists $v \in \hat{Z}i$ such that $\bar{\gamma}^T \cdot u = 0$ for any $\gamma \in N_{\tilde{a}}$, where $u = v-e$. Then, by (vii) in Lemma 6.1.2, $\hat{X}^* u = \bar{X}^T \cdot u = 0$. Now, if $\Gamma \in \mathcal{K}^{c_2}$ and $d^T \cdot \Gamma = 0$, i.e., $\Gamma \in \mathfrak{N}$ by (6.1.22), then, by Lemma 6.1.3, there exists $\omega \in \mathcal{K}^{c_o}$ such that $\Gamma = X \cdot \omega$. Hence, we have $0 = \omega^T \cdot (X^T \cdot u) = (\omega^T \cdot X^T) \cdot u = (X \cdot \omega)^T \cdot u = \Gamma^T \cdot u$, and (i) is proved.

Conversely, let the hypothesis in (ii) be satisfied. Then clearly i satisfies $\tilde{K}_2$. Letting again $u = v-e$, we have for any $\omega \in \mathcal{K}^{c_o}$ by Lemma 6.1.3, $0 = (X \cdot \omega)^T \cdot u = (\omega^T \cdot X^T) \cdot u = \omega^T \cdot (X^T \cdot u)$. Denote $X^T \cdot u = [z_k]$, and let $\nu \in \mathcal{K}$. Putting $\omega = [0, 0, \ldots, \nu, 0, \ldots, 0]^T$ with ν standing at k-th place, we infer from the above relation that $0 = \nu z_k$; thus, by completeness of $\mathcal{K}$, $z_k = 0$ for $k = 1, 2, \ldots, c_o$. Consequently, $X^T \cdot u = 0$, i.e., $u \in N_{\hat{X}^*}$; hence, by (vii) in Lemma 6.2.2, i satisfies $\tilde{K}_1$ which completes the proof.

Remark 2. (a) The proposition (i) in Theorem 6.1.5 is a generalization of the classical Tellegen's theorem, and (ii) furnishes its converse.

(b) Assume that $\hat{n} = (\hat{Z}, G)$ is an algebraic network whose underlying space L happens to be a Hilbert space; then Theorem 2.2.1, which is valid for Hilbert (not necessarily finite) networks, follows for $\hat{n}$ from Theorem 6.1.5. To see this, let $L' = R^1$ and let $\mathcal{K} = \{A_\alpha : \alpha \in L\}$, where each operator $A_\alpha : L \to R^1$ is defined by $A_\alpha x = \langle \alpha, x \rangle$. Clearly, $\mathcal{K}$ is complete, since $\langle \alpha, x \rangle = 0$ for all $\alpha \in L$ implies that $x = 0$. Moreover, it is easy to see that

$$\Gamma = [A_{\alpha_1}, A_{\alpha_2}, \ldots, A_{\alpha_{c_2}}]^T \in \mathcal{K}^{c_2} \text{ and } d^T \cdot \Gamma = 0 \Leftrightarrow$$

$$\alpha = [\alpha_1, \alpha_2, \ldots, \alpha_{c_2}]^T \in L^{c_2} \text{ and } d^T \cdot \alpha = 0.$$

Thus, with $v-e = u = [u_1, u_2, \ldots, u_{c_2}]^T$, (6.1.23) reads

$$0 = \Gamma^T \cdot u = \sum_{j=1}^{c_2} A_{\alpha_j} u_j = \sum_{j=1}^{c_2} \langle \alpha_j, u_j \rangle = \langle \alpha, u \rangle_{c_2}.$$

This, however, is precisely condition K_1. Hence, our claim.

(c) A special case of the space $\mathcal{K}$, and consequently, a special case of Theorem 6.1.5 is obtained if L is a (not necessarily commutative) algebra, i.e., if to each pair $\ell_1, \ell_2 \in L$ there is assigned an element $\ell_1 \ell_2 \in L$ such that the following requirements are satisfied:

(1) $\ell_1(\ell_2\ell_3) = (\ell_1\ell_2)\ell_3$,

(2) $\ell_1(\ell_2+\ell_3) = \ell_1\ell_2 + \ell_1\ell_3$ and $(\ell_1+\ell_2)\ell_3 = \ell_1\ell_3 + \ell_2\ell_3$,

(3) $\alpha(\ell_1\ell_2) = (\alpha\ell_1)\ell_2 = \ell_1(\alpha\ell_2)$ for every scalar α.

Indeed, in this case we let $L' = L$, and for $\mathcal{K}$ we take the set of all operators (ℓ), defined by: $\ell \in L$, $(\ell) : L \to L$, $(\ell)x = \ell x$. Then $\mathcal{K}$ is clearly a linear subspace of (L,L). Note also that if L contains an identity, i.e., an element $1 \in L$ such that $1x = x1 = x$ for every $x \in L$, then clearly $\mathcal{K}$ contains a 1-to-1 operator and we can apply proposition (ii).

We now establish an equivalent definition of a solution of an algebraic network.

Theorem 6.1.6. Let $\hat{n} = (\hat{Z},G)$ be an algebraic network.

(i) If, for some $e \in L^{c_2}$, $\hat{n}$ possesses a solution $i \in L^{c_2}$, then there exists $u \in \hat{Z}i$ and $v \in L^{c_1}$ such that

$$u + \hat{a}^*v = e. \tag{6.1.24}$$

(ii) If, for some $e \in L^{c_2}$, there exists elements $i \in N_{\hat{a}} \cap D$, $u \in \hat{Z}i$ and $v \in L^{c_1}$ such that (6.1.24) holds, then i is a solution of $\hat{n}$ corresponding to e.

Proof: (i) Assume that $i \in L^{c_2}$ is a solution of $\hat{n}$ corresponding to e; then, by $\tilde{K}_2$, $i \in N_{\hat{a}} \cap D$, and, by $\tilde{K}_1$, there exists $u \in \hat{Z}i$ such that $u-e \in N_{\hat{X}^*}$ (see Lemma 6.1.2 (vii)). However, by Lemma 6.1.2 (iv), $N_{\hat{X}^*} = \hat{a}^*L^{c_1}$, i.e., there exists an element $-v \in L^{c_1}$ such that $u-e = \hat{a}^*(-v)$; this, however, is equation (6.1.24).

(ii) Conversely, if there exists $i \in N_{\hat{a}} \cap D$, $u \in \hat{Z}i$ and $v \in L^{c_1}$ such that (6.1.24) holds, then $\tilde{K}_2$ is satisfied, and we have by Lemma 6.1.2 (iv), $u-e = \hat{a}^*(-v) \in \hat{a}^*L^{c_1} = N_{\hat{X}^*}$; hence, i satisfies $\tilde{K}_1$, and consequently, i is a solution of $\hat{n}$ corresponding to e.

Following the ideas developed in Section 1.4, we could prove theorems analogous to Theorems 1.4.2-1.4.4. Since this is a quite straightforward process, we will omit the details and, instead, we will prove a theorem which appears as a modification of Theorem 1.4.4. To this end, however, we have to make some considerations concerning a finite oriented graph.

Let G be a finite oriented graph having the set of branches $\mathcal{B}$ and the set of nodes $\mathcal{V}$. If $b_i \in \mathcal{B}$, let $v_{m_i} \in \mathcal{V}$ and $v_{m_i'} \in \mathcal{V}$ be the terminal and initial node of b_i, respectively, i.e., $d_{i,m_i} = 1$, $d_{i,m_i'} = -1$, $d_{i,j} = 0$ for

$j \neq m_i, m_i'$ (see (2.1.3)). If δ_i is either 1 or -1, let

$$\partial(\delta_i b_i) = (v_{m_i'}, v_{m_i}) \text{ for } \delta_i = 1,$$

and

$$\partial(\delta_i b_i) = (v_{m_i}, v_{m_i'}) \text{ for } \delta_i = -1.$$

Now, let $v_\alpha, v_\beta \in \mathfrak{V}$; we will say that v_α is related to v_β, if there exists a sequence $b_{i_1}, b_{i_2}, \ldots, b_{i_m} \in \mathfrak{B}$ and numbers $\delta_{i_1}, \delta_{i_2}, \ldots, \delta_{i_m}$, each having value 1 or -1, such that $\partial(\delta_{i_1} b_{i_1}) = (v_\alpha, v_\gamma)$, $\partial(\delta_{i_m} b_{i_m}) = (v_\varepsilon, v_\beta)$ and $\partial(\delta_{i_k} b_{i_k}) = (v_{r_k}, v_{r_k'})$ with $v_{r_k'} = v_{r_{k+1}}$ for $k = 1,2,\ldots,m-1$. Thus, crudely speaking, v_α is related to v_β, if we can get from v_α to v_β by following a certain sequence of branches.

It is clear that v_α is related to v_β, if and only if v_β is related to v_α.

The graph G will be called connected, if for any $v_\alpha, v_\beta \in \mathfrak{V}$, v_α is related to v_β.

From the definition of the matrix d it follows easily that, if G is connected, every solution $z \in L^{c_1}$ of the equation $d \cdot z = 0$ has the form $z = [\ell, \ell, \ldots, \ell]^T$, where ℓ is some element in L. Indeed, if nodes v_α and v_β are connected by some branch b_i, and if $z = [z_1, z_2, \ldots, z_{c_1}]^T$, it follows that $z_\alpha = z_\beta$; hence, by connectedness of G, we must have $z_1 = z_2 = \ldots = z_{c_1}$.

Let d' be the $c_2 \times (c_1 - 1)$ matrix obtained from d by deleting the last column.

Then from Theorem 6.1.6 we obtain the following result:

<u>Proposition</u>. Let $\hat{\mathfrak{N}} = (\hat{Z}, G)$ be an algebraic network, let $\hat{Z}$ be an operator defined on L^{c_2} and let G be connected.

(i)* If, for some $e \in L^{c_2}$, $\hat{\mathfrak{N}}$ possesses a solution $i \in L^{c_2}$, then there exists a unique element $v' \in L^{c_1 - 1}$ such that

$$\hat{Z}i + d' \cdot v' = e. \qquad (6.1.25)$$

(ii)* If, for some $e \in L^{c_2}$, there exists elements $i \in N_{\hat{a}}$ and $v' \in L^{c_1 - 1}$ such that (6.1.25) holds, then i is a solution of $\hat{\mathfrak{N}}$ corresponding to e.

<u>Proof</u>: (i)* If $i \in L^{c_2}$ is a solution of $\hat{\mathfrak{N}}$ corresponding to $e \in L^{c_2}$, then, by Theorem 6.1.6 (i), there exists $v \in L^{c_1}$ such that

$$\hat{Z}i + d \cdot v = e. \tag{6.1.26}$$

Since by connectedness of G every solution $z \in L^{c_1}$ of $d \cdot z = 0$ has the form $z^* = [\ell, \ell, \ldots, \ell]^T$, $\ell \in L$, let us put $\tilde{v} = [v_{c_1}, v_{c_1}, \ldots, v_{c_1}]^T$, where v_{c_1} is the c_1-th component of v; then $d \cdot \tilde{v} = 0$ and (6.1.26) yields

$$\hat{Z}i + d \cdot (v - \tilde{v}) = e. \tag{6.1.27}$$

However, $d \cdot (v - \tilde{v}) = d' \cdot v'$ with $v' \in L^{c_1 - 1}$, $v' = [v_1 - v_{c_1}, v_2 - v_{c_1}, \ldots, v_{c_1 - 1} - v_{c_1}, 0]^T$, and v' is clearly determined uniquely.

(ii)* Clearly, if (6.1.25) holds, then, putting $v = [v', 0]^T$, $v \in L^{c_1}$, we have $d' \cdot v' = d \cdot v$ and (6.1.25) becomes (6.1.24) in Theorem 6.1.6 (ii), which concludes the proof.

Lemma 6.1.4. Let G be a connected graph, and let d' have the meaning defined above. Then for any $y \in L^{c_1 - 1}$ there exists a $z \in L^{c_2}$ such that $d'^T \cdot z = y$.

Proof: Since every solution $z \in L^{c_1}$ of $d \cdot z = 0$ has the form $z = [\ell, \ell, \ldots, \ell]^T$, it follows that rank $d = c_1 - 1$ (it suffices to set $L = R^1$ for a while). Moreover, since the sum of all columns of d is zero, the $c_2 \times (c_1 - 1)$ matrix d' has also rank $c_1 - 1$. Thus, if $\eta \in R^{c_1 - 1}$ is arbitrary, then there always exists $\xi \in R^{c_2}$ such that $d'^T \cdot \xi = \eta$, because clearly rank $[d'^T \vdots \eta]$ = rank d'^T.

Now, let $y \in L^{c_1 - 1}$. If $\mathfrak{B}$ is some basis in L, then there exists elements $\ell_1, \ell_2, \ldots, \ell_m \in \mathfrak{B}$ and vectors $\eta_1, \eta_2, \ldots, \eta_m \in R^{c_1 - 1}$ such that

$$y = \sum_{j=1}^{m} \eta_j \ell_j. \tag{6.1.28}$$

However, by the above there exists $\xi_1, \xi_2, \ldots, \xi_m \in R^{c_2}$ such that $d'^T \cdot \xi_j = \eta_j$, $j = 1, 2, \ldots, m$. Putting $z = \sum_{j=1}^{m} \xi_j \ell_j$, we have $z \in L^{c_2}$, and by (6.1.28),

$$d'^T \cdot z = \sum_{j=1}^{m} d'^T \cdot \xi_j \ell_j = \sum_{j=1}^{m} \eta_j \ell_j = y.$$

Now, we are ready to prove the theorem.

Theorem 6.1.7. Let $\hat{n} = (\hat{Z}, G)$ be an algebraic network, let G be connected and let d' have the meaning introduced above. Moreover, assume that $\hat{Z}$ is a linear operator defined on L^{c_2} and represented by a $c_2 \times c_2$ matrix Z, i.e., $\hat{Z}x = Z \cdot x$, where the elements Z_{ik} of Z are linear operators $L \to L$. Define the operator $\hat{M} : L^{c_2 + c_1 - 1} \to L^{c_2 + c_1 - 1}$ by $\hat{M}z = M \cdot z$, where

$$M = \left[\begin{array}{c|c} Z & d' \\ \hline d'^T & 0 \end{array}\right] \qquad (6.1.29)$$

Then

(i) $Q(L^{c_2}) = L^{c_2}$, if and only if $\hat{M}$ maps $L^{c_2+c_1-1}$ onto $L^{c_2+c_1-1}$,

(ii) $\hat{n}$ is regular on L^{c_2}, if and only if $\hat{M}$ is 1-to-1.

Proof: (i) Let $\hat{M}$ map $L^{c_2+c_1-1}$ onto itself. Choose $e \in L^{c_2}$ and let $e' \in L^{c_2+c_1-1}$ be defined by $e' = [e^T \vdots 0]^T$; then, by our hypothesis, there exists vectors $i \in L^{c_2}$ and $v' \in L^{c_1-1}$ such that the vector $w \in L^{c_2+c_1-1}$, defined by $w = [i^T \vdots v'^T]^T$ satisfies the equality $\hat{M}w = e'$, i.e., by definition of $\hat{M}$,

$$Z \cdot i + d' \cdot v' = e, \qquad (6.1.30)$$

and

$$d'^T \cdot i = 0. \qquad (6.1.31)$$

However, since the sum of all rows of the matrix d^T is zero, it follows that equation (6.1.31) is equivalent to $d^T \cdot i = 0$, i.e., $i \in N_{\hat{a}}$; hence, by our Proposition (ii)* ((6.1.30) is the same as (6.1.25) by definition of Z), i is a solution of $\hat{n}$ corresponding to e. Thus, $Q(L^{c_2}) = L^{c_2}$.

Conversely, assume that $Q(L^{c_2}) = L^{c_2}$. Choose $e \in L^{c_2}$, $y \in L^{c_1-1}$, and let $e' \in L^{c_2+c_1-1}$ be defined by $e' = [e^T \vdots y^T]^T$. Then, by Lemma 6.1.4, there exists a vector $z \in L^{c_2}$ such that $d'^T \cdot z = y$. Define $\tilde{e} \in L^{c_2}$ by $\tilde{e} = e - Z \cdot z$; by our hypothesis, there exists $i \in L^{c_2}$ which is a solution of $\hat{n}$ corresponding to $\tilde{e}$, i.e., by our Proposition (i)*, there exists a unique $v' \in L^{c_1-1}$ such that

$$Z \cdot i + d' \cdot v' = e - Z \cdot z \qquad (6.1.32)$$

and $d^T \cdot i = 0$. However, since $d^T \cdot i = 0$ is equivalent to $d'^T \cdot i = 0$, we have by the above $d'^T \cdot (i+z) = y$. Thus, we can write by (6.1.32) using linearity of Z,

$$Z \cdot (i+z) + d' \cdot v' = e, \quad d'^T \cdot (i+z) = y;$$

these equations mean that $\hat{M}w = e'$, where $w = [i^T + z^T \vdots v'^T]^T$. Hence, $\hat{M}$ maps $L^{c_2+c_1-1}$ onto itself, and our claim (i) is proven.

(ii) Assume first that $\hat{M}$ is 1-to-1. Suppose that, for some $e \in L^{c_2}$, $i_j \in N_{\hat{a}}$, $j = 1,2$ is a solution of $\hat{n}$ corresponding to e, i.e, by Proposition (i)*, there exists $v_j' \in L^{c_1-1}$ such that

$$Z\cdot i_j + d'\cdot v_j' = e. \tag{6.1.33}$$

Since $i_j \in N_{\hat{a}}$ is equivalent to $d'^T\cdot i_j = 0$, (6.1.33) and linearity of Z imply that

$$Z\cdot(i_1-i_2) + d'\cdot(v_1'-v_2') = 0, \quad d'^T\cdot(i_1-i_2) = 0, \tag{6.1.34}$$

i.e., $\hat{M}w_o = 0$ with $w_o = [(i_1-i_2)^T \,\vdots\, (v_1-v_2)^T]^T$; thus, by our hypothesis, $w_o = 0 \Rightarrow i_1 = i_2$. Hence, $\hat{n}$ is regular on L^{c_2}.

Conversely, assume that $\hat{n}$ is regular on L^{c_2}. Suppose that, for some $w_o \in L^{c_2+c_1-1}$, $w_o = [i_o^T \,\vdots\, v_o'^T]^T$, $i_o \in L^{c_2}$, $v_o' \in L^{c_1-1}$, we have $\hat{M}w_o = 0$, i.e.,

$$Z\cdot i_o + d'\cdot v_o' = 0, \; d'^T\cdot i_o = 0. \tag{6.1.35}$$

This, however, means according to Proposition (ii)* that i_o is a solution of $\hat{n}$ corresponding to $0 \in L^{c_2}$, i.e., by our hypothesis and linearity of Z, $i_o = 0$. On the other hand, by Proposition (i)*, the element v_o' in (6.1.35) is determined uniquely; since $i_o = 0$ implies by (6.1.35) that $d'\cdot v_o' = 0$, we have necessarily $v_o' = 0$. Hence, $\hat{M}w_o = 0 \Rightarrow w_o = 0$, i.e., $\hat{M}$ is 1-to-1 which concludes the proof.

Remark 3. Theorem 6.1.7 becomes particularly useful, if the operators Z_{ik} constituting the matrix Z belong to some algebraic field. In this case, we clearly have: $\hat{n}$ is regular on L^{c_2} and $Q(L^{c_2}) = L^{c_2} \Leftrightarrow \det M \neq 0$.

A similar conclusion can be drawn from Theorem 6.1.2, 6.1.3 and 6.1.4. Indeed, we have:

Let $\hat{n} = (\hat{Z},G)$ be an algebraic network, and let $\hat{Z}$ be a linear operator defined on L^{c_2}; if the elements of the matrix Z representing $\hat{Z}$ belong to some algebraic field, then

$$\hat{n} \text{ is regular on } L^{c_2} \text{ and } Q(L^{c_2}) = L^{c_2} \Leftrightarrow \det(\bar{X}^T\cdot Z\cdot X) \neq 0.$$

In this case, the admittance A of $\hat{n}$ is represented by the matrix $A' = X\cdot(\bar{X}^T\cdot Z\cdot X)^{-1}\cdot\bar{X}^T$.

The proof of this assertion is obvious. An important situation with a network of this kind will be discussed in the next section.

Problems.

6.1.1 Let us modify Problem 3.5.1 as follows: Let now G be a finite graph, and let $L_2[0,\tau]$ be replaced by the space L consisting of all locally integrable functions on $[0,\infty)$. Also, let the assumptions (i)-(iii) be changed accordingly. Using Theorem 6.1.5, prove the proposition:

Let $e \in L^{c_2}$; then i is an elementary solution of $\hat{n}$ corresponding to $e \Leftrightarrow i \in D$ and i is a generalized solution of $\hat{n}$ corresponding to e.

6.1.2 In the previous problem, let L be replaced by the space B consisting of all locally bounded, measurable functions on $[0,\infty)$. Prove:

Let $e \in B^{c_2}$; then i is an elementary solution of $\hat{n}$ corresponding to $e \Leftrightarrow i \in D$, $d^T \cdot i = 0$ and

$$\int_0^t c^T(\tau)\cdot[(\hat{Z}i)(\tau)-e(\tau)]d\tau = 0$$

for all $c \in B^{c_2}$ with $d^T \cdot c = 0$.

6.2 Finite L,R,C-Networks With Constant Elements

In this section, we will briefly discuss finite L,R,C-networks with constant elements when the underlying space L is the space of distributions vanishing on the negative semiaxis.

Let $\mathcal{D}$ be the space of all complex-valued, infinitely smooth functions on R^1 which have compact support. Moreover, let $\mathcal{D}'_+$ be the space of all Schwartz distributions vanishing on $(-\infty,0)$, (see [13] or [14]). Note that if a(t) is infinitely smooth on R^1 and $x \in \mathcal{D}'_+$, then $ax \in \mathcal{D}'_+$.

If $x \in \mathcal{D}'_+$, let $x^{(-1)} : \mathcal{D} \to R^1$ be defined by

$$(x^{(-1)},\varphi) = (x,\varphi^{[-1]}), \tag{6.2.1}$$

where

$$\varphi^{[-1]}(t) = -\int_{-\infty}^t \varphi(\tau)d\tau + \left(\int_{-\infty}^{\infty} \varphi(\tau)d\tau\right)\cdot\int_{-\infty}^t \varphi_0(\tau)d\tau$$

and $\varphi_0 \in \mathcal{D}$ is such that supp $\varphi_0 \subset (-\infty,0)$ and $\int_{-\infty}^{\infty} \varphi_0(\tau)d\tau = 1$.

We can easily verify that the following is true [13]:

(1) $x^{(-1)} \in \mathcal{D}'_+$ and $x^{(-1)}$ does not depend on the choice of φ_0.

(2) If $x \in \mathcal{D}'_+$ is regular and corresponds to $\tilde{x}(t)$, then $x^{(-1)}$ is regular and corresponds to $\int_0^t \tilde{x}(\tau)d\tau$.

(3) $(\alpha_1x_1+\alpha_2x_2)^{(-1)} = \alpha_1x_1^{(-1)} + \alpha_2x_2^{(-1)}$ for any constants α_1,α_2 and $x_1,x_2 \in \mathcal{D}'_+$.

(4) $(x')^{(-1)} = (x^{(-1)})' = x$ for each $x \in \mathcal{D}'_+$.

Next, let I be the identity operator on $\mathcal{D}'_+$, and let $D : \mathcal{D}'_+ \to \mathcal{D}'_+$ be defined by $Dx = x'$. Then we have:

<u>Lemma 6.2.1.</u> For every number α, the operator $D + \alpha I$ is 1-to-1 from $\mathcal{D}'_+$ onto $\mathcal{D}'_+$; moreover, for each $x \in \mathcal{D}'_+$,

$$(D+\alpha I)^{-1}x = e^{-\alpha t}(e^{\alpha t}x)^{(-1)}. \qquad (6.2.2)$$

The proof follows readily from (4) and the obvious formula $(ax)' = a'x + ax'$, $x \in \mathcal{D}'_+$, $a(t)$ infinitely smooth.

<u>Lemma 6.2.2.</u> Let $\mathfrak{R}$ be the linear span of the set $\{I,D,D^2,D^3,\ldots\}$; then

(i) $\mathfrak{R}$ is a commutative ring,

(ii) $\{I,D,D^2,D^3,\ldots\}$ is a basis in $\mathfrak{R}$,

(iii) if $Q \in \mathfrak{R}$, $Q \neq 0$, then Q is 1-to-1 from $\mathcal{D}'_+$ onto $\mathcal{D}'_+$.

<u>Proof</u>: The proposition (i) is an obvious consequence of linearity of the operator D. To prove (ii), suppose that $\{I,D,D^2,\ldots\}$ is not a basis of $\mathfrak{R}$, i.e., the set $\{I,D,D^2,\ldots\}$ is linearly dependent. Then there exists $n \geq 0$ and numbers $a_0,a_1,\ldots,a_n$ with $a_n \neq 0$ such that $P(D) = a_nD^n + a_{n-1}D^{n-1} + \ldots$ $\ldots + a_0I = 0$. Choose $x \in \mathcal{D}'_+$, $x \neq 0$. If $P(\xi) = a_n \prod_{i=1}^{n} (\xi-\alpha_i)$, we have by (i), $P(D)x = (a_n \prod_{i=1}^{n} (D-\alpha_iI))x = 0$. Invoking Lemma 6.2.1 and applying successively operators $(D-\alpha_iI)^{-1}$, $i = 1,2,\ldots,n$ to this equation, it follows that $a_nx = 0$, i.e., $a_n = 0$ - contradiction.

As for (iii), it suffices to realize that, if $Q \neq 0$, then $Q = a_n \prod_{i=1}^{n} (D-\alpha_iI)$ for some integer $n \geq 0$ and $a_n \neq 0$; since each $D-\alpha_iI$ is 1-to-1 from $\mathcal{D}'_+$ onto $\mathcal{D}'_+$ by Lemma 6.2.1, so is Q. Hence, the proof.

Now, define:

$$\mathfrak{A} = \{R : R = Q^{-1}P,\ P,Q \in \mathfrak{R},\ Q \neq 0\}. \qquad (6.2.3)$$

Then we have:

<u>Lemma 6.2.3.</u> The set $\mathfrak{A}$ is a field.

<u>Proof</u>: If $P_1,P_2 \in \mathfrak{R}$, then by Lemma 6.2.2 (i), $P_1P_2 = P_2P_1$. If $P_1 \neq 0$, it follows readily that

$$P_1^{-1}P_2 = P_2P_1^{-1}. \qquad (6.2.4)$$

Moreover, if, in addition, $P_2 \neq 0$, we have $P_1^{-1}P_2^{-1} = P_2^{-1}P_1^{-1}$. Using these equalities we easily confirm that, for $R_1, R_2 \in \mathfrak{A}$, $R_1R_2 = R_2R_1$, $R_1R_2 \in \mathfrak{N}$, $R_1 + R_2 \in \mathfrak{A}$. Hence, $\mathfrak{A}$ is a field.

Note that, due to (6.2.4), every operator $R \in \mathfrak{N}$ can be written as a "fraction P/Q"; also, the partial fraction expansion is true, i.e., if $R \in \mathfrak{N}$, then there exists uniquely determined numbers a_j, $j = 1,2,\ldots,n$, α_i, $i = 1,2,\ldots,m$ and λ_{ik}, $i = 1,2,\ldots,m$, $k = 1,2,\ldots,q$ such that

$$R = \sum_{j=0}^{n} a_j D^j + \sum_{i=1}^{m} \sum_{k=1}^{q} \lambda_{ik}\{(D-\alpha_i I)^{-1}\}^k; \tag{6.2.5}$$

here, we understand that $D^o = I$. Consequently, if $x \in \mathcal{D}'_+$ is given, Rx can easily be established by using the decomposition (6.2.5) and formula (6.2.2).

Let us now apply these facts to a finite L,R,C-network.

Let G be a finite oriented graph having c_2 branches ($c_2 < \aleph_o$), and let L,R,S be $c_2 \times c_2$ matrices with constant elements; furthermore, let $\hat{Z} : \mathcal{D}'^{c_2}_+ \to \mathcal{D}'^{c_2}_+$ be defined by $\hat{Z}x = Z\cdot x$, where

$$Z = LD + R + SD^{-1}. \tag{6.2.6}$$

Then the algebraic network $\hat{n} = (\hat{Z},G)$ will be called a finite L,R,C-network with constant elements.

Using the results obtained in Section 6.1 and Lemma 6.2.3, we get immediately the following proposition:

<u>Theorem 6.2.1.</u> Let $\hat{n}$ be a finite L,R,C-network with constant elements; then $\hat{n}$ is regular on $\mathcal{D}'^{c_2}_+$ and $Q(\mathcal{D}'^{c_2}_+) = \mathcal{D}'^{c_2}_+$, if and only if $\det(\bar{X}^T\cdot Z\circ X) \neq 0$. In this case, the admittance A of $\hat{n}$ is generated by the matrix

$$A' = X\cdot(\bar{X}^T\cdot Z\cdot X)^{-1}\cdot\bar{X}^T. \tag{6.2.7}$$

An analogous result follows from Theorem 6.1.7, but we omit the details.

We conclude this section with several remarks:

1. The network model we discussed above described the physical situation of an L,R,C-network being in an equilibrium state at $t = 0$, i.e., the initial currents are zero and there are no charges on capacitors. The more general case, however, when the network is not at rest at $t = 0$, can easily be reduced to the previous one. Indeed, if $i_o \in R^{c_2}$ and $q_o \in R^{c_2}$ describe the initial currents and charges, respectively, then the operator $\hat{Z}$ is defined on $\mathcal{D}'^{c_2}_+$ by

$$\hat{Z}x = L\cdot(x'-i_o\delta_o) + R\cdot x + S\cdot(x^{(-1)} + q_oH_o), \qquad (6.2.8)$$

where δ_o is the Dirac distribution and $H_o = \delta_o^{(-1)}$. (For more details about this description see [13]). From (6.2.8) it readily follows that a solution i of such a network corresponding to some $e \in \mathcal{D}_+'^{c_2}$ coincides with the solution of the L,R,C-network $\hat{n}$ corresponding to $e + L\cdot i_o\delta_o - S\cdot q_oH_o$. Hence, introducing $\hat{Z}$ by (6.2.6) entails no loss of generality.

2. The reason for choosing $\mathcal{D}_+'$ as the underlying space for an L,R,C-network is given by the fact that the distributions are infinitely differentiable, and consequently, the requirements guaranteeing the existence of a solution are minimal. Also, this approach generalizes the case when currents and voltages are described by locally integrable functions, or possibly by differentiable functions. To see the power of this approach, consider the network sketched in Figure 6.2.

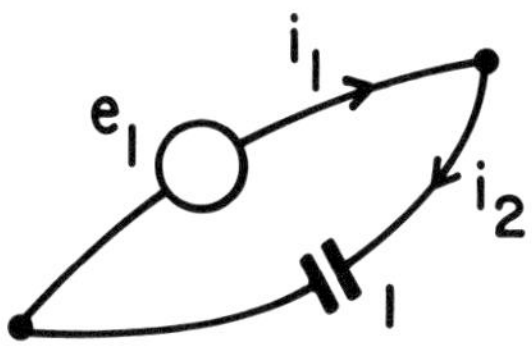

Figure 6.2

Here, $L = R = 0$, $S = \text{diag}(0,1)$, $e = [1,0]^T$, and the network is assumed to be at rest at $t = 0$. If we are looking for a solution $[i_1,i_2]^T$ in the space of locally integrable functions defined on $[0,\infty)$, then we have by $\tilde{K}_1$, $\tilde{K}_2$, $\int_0^t i_1(\tau)d\tau = 1$ for all $t > 0$. However, there clearly does not exist i_1 with this property. On the other hand, if we are looking for a solution in $\mathcal{D}_+'^{c_2}$, then the above equation is replaced by $i_1^{(-1)} = H_o$, and we confirm easily that the vector $[\delta_o,\delta_o]^T$ is the corresponding solution. A more detailed treatment of these problems can be found in [13], [16].

3. It is not difficult to apply results in Section 6.1 to L,R,C-networks having $\mathcal{D}_+'$ as the underlying space, but whose elements are described by smooth functions on $[0,\infty)$. Note that in this case conditions for the existence and uniqueness of a solution are not so simple as those in Theorem 6.2.1. Among other things, we have to make the assumption that rank $L(t)$ is constant in $[0,\infty)$. The details can be found in [13], [15].

Problems.

6.2.1 Let $x_n = [x_k^n] \in \mathcal{D}_+'^{c_2}$, $n = 0,1,2,\ldots$; we will write $x_n \to x_o$, if $x_k^n \to x_k^o$ in $\mathcal{D}_+'$ for $k = 1,2,\ldots,c_2$, (see [14]). Let $\hat{\eta}$ be a finite L,R,C-network with constant elements which satisfies the condition $\det(\bar{X}^T \cdot Z \cdot X) \neq 0$, let $e_n \in \mathcal{D}_+'^{c_2}$ and let i_n be the solution of $\hat{\eta}$ corresponding to e_n, $n = 0,1,2,\ldots$; if $e_n \to e_o$, then $i_n \to i_o$.

SOLUTIONS TO PROBLEMS

Section 1.1

1.1.1 Choose $x_o \in X$; then by (i), $A0 = A(0x_o) = 0Ax_o = \{0\}$. Next, let $x \in X$ and suppose that $y_1, y_2 \in Ax$; then, by (ii), $\{0\} = A0 = A(x-x) = Ax+A(-x) = Ax+(-1)Ax$. Since $y_1 \in Ax$ and $-y_2 \in (-1)Ax$, we have $y_1-y_2 \in Ax + (-1)Ax = \{0\}$. Hence, $y_1-y_2 = 0$.

1.1.2 Choose a point $x_o \in D$, and let $x' \in D$ be an arbitrary point, $x' \neq x_o$. By connectedness, there exists $x_1, x_2, \ldots, x_{n-1} \in D$ such that $[x_k, x_{k+1}] \subset D$ for $k = 0,1,\ldots,n-1$.

For any $\lambda \in [0,1]$, let

$$\varphi(\lambda) = \|A(\lambda x_o + (1-\lambda)x_1) - Ax_o\|. \qquad (\ast\ast)$$

If $\lambda, \lambda' \in [0,1]$, we have by $(*)$ and $(\ast\ast)$,

$$|\varphi(\lambda) - \varphi(\lambda')| \leq \alpha|\lambda-\lambda'|^p \|x_o-x_1\|^p.$$

Hence, for $\lambda \neq \lambda'$,

$$|\lambda-\lambda'|^{-1} \cdot |\varphi(\lambda) - \varphi(\lambda')| \leq \alpha|\lambda-\lambda'|^{p-1}\|x_o-x_1\|^p,$$

i.e., φ possesses a derivative on $[0,1]$ and $\varphi'(\lambda) = 0$ for each $\lambda \in [0,1]$. Thus, $\varphi = c$ on $[0,1]$ and by $(\ast\ast)$, $\varphi(1) = 0$. Consequently, $\varphi = 0$, so that, for $\lambda = 0$, $Ax_1 = Ax_o$. Repeating the argument it follows that $Ax' = Ax_o$.

1.1.3 For $x_1, x_2 \in K$ we have $\langle Ax_1-Ax_2, x_1-x_2\rangle = \int_0^T (x_1'-x_2')(x_1-x_2)dt = \frac{1}{2}(x_1-x_2)^2\Big|_0^T = \frac{1}{2}(x_1(T)-x_2(T))^2 \geq 0$.

Section 1.2

1.2.1 Clearly, $P(x_1,x_2) = (x_1-x_2,0)$, and on $N \cap D = N$ we have $Z(t,0) = (t^3,-t)$; hence, $PZ(t,0) = (t^3+t,0)$. Let $\varphi(t) = t^3+t$; then φ is 1-to-1 from R^1 onto R^1. Consequently, PZ is 1-to-1 from N onto N, and $(PZ)^{-1}(\sigma,0) = (\varphi^{-1}(\sigma),0)$. Thus, $\mathfrak{h}$ is regular on R^2 and $Q(R^2) = R^2$. Also, $A(x_1,x_2) = (PZ)^{-1}P(x_1,x_2) = (\varphi^{-1}(x_1-x_2),0)$.

1.2.2 No. To see it, consider $\mathfrak{h} = [Z,N,M]$ with $\mathfrak{X} = R^2$, $N = \{(t,0) : t \in R^1\}$, $M = \{(0,\tau) : \tau \in R^1\}$, $D = R^2$ and let $Z(t,\tau) = (t,0)$ for $t \neq 0$,

$Z(0,\tau) = \{(0,\xi) : \xi \in [0,1]\}$. Then clearly $P(t,\tau) = (t,0)$, and on $N \cap D = N$, $PZ(t,0) = (t,0)$ for all $t \in R^1$. Hence, h is regular on R^2 and $Q(R^2) = R^2$. Also, $A(t,\tau) = (PZ)^{-1}P(t,\tau) = (t,0)$. Now, since $e = (0,0) \in Q(R^2)$, by regularity there exists a unique solution i corresponding to e, and we have $i = (0,0)$. By K_1, there exists $v \in Zi$ such that $v-e = v \in M$. However, $Zi = \{(0,\xi) : \xi \in [0,1]\}$, and consequently, there exists infinitely many v's which satisfy the relation $v \in M$.

1.2.3 (i) and (ii) is an elementary consequence of (1.2.2) and Theorem 1.2.2, respectively. As for (iii), assume

(a) that $N \cap R = \{0\}$. Then $x \in N \cap D$ and $PZx = 0 \Rightarrow x \in N$, $x \in D$ and $Zx \in M \Rightarrow x \in N \cap R \Rightarrow x = 0$. Thus, PZ is 1-to-1 on $N \cap D \Rightarrow h$ is regular on D.

(b) Conversely, let h be regular on D, i.e., PZ is 1-to-1 on $N \cap D$. Then $x \in N \cap R \Rightarrow x \in N$, $x \in D$ and $Zx \in M \Rightarrow x \in N \cap D$ and $PZx = 0 \Rightarrow x = 0$. Hence, $N \cap R = \{0\}$.

Section 1.3

1.3.1 By (i), $N_a \cap D = N_a$, and by (ii), PZ is 1-to-1 from N_a onto N_a. Since PZ is linear and bounded on N_a, and N_a is a B-space, it follows by the open-mapping theorem that $(PZ)^{-1}$ is bounded on N_a. Hence, $A = (PZ)^{-1}P \in [\mathcal{H},\mathcal{H}]$.

1.3.2 Condition (ii) implies readily that Z is linear. Moreover, using (i), (ii) and considering the inequality $\langle Z(x+\lambda y),x+\lambda y\rangle \geq 0$ it follows by a standard argument that, for any $x \in \mathcal{H}$, $\langle Zx,x\rangle = 0 \Leftrightarrow Zx = 0$.

(a) Assume that $N_a \cap N_z = \{0\}$. If, for some $e \in \mathcal{H}$, h has solutions i_1,i_2, then $i_1,i_2 \in N_a$ and Zi_1-e, $Zi_2-e \in N_a^\perp \Rightarrow i_1-i_2 \in N_a$ and $Z(i_1-i_2) \in N_a^\perp \Rightarrow \langle Z(i_1-i_2),i_1-i_2\rangle = 0 \Rightarrow Z(i_1-i_2) = 0 \Rightarrow i_1-i_2 \in N_z \Rightarrow i_1-i_2 = 0$. Thus, h is regular on $\mathcal{H}$.

(b) Assume that h is regular on $\mathcal{H}$. Then $i = 0$ is the unique solution corresponding to $e = 0$. If $x \in N_a \cap N_z$, then $x \in N_a$ and $Zx = 0 \Rightarrow x$ is a solution corresponding to $e = 0$ (we have $Zx-e = 0 \in N_a^\perp$) $\Rightarrow x = 0$. Hence, $N_a \cap N_z = \{0\}$.

1.3.3 First observe that the following proposition is true:

Let $X \neq \emptyset$, let $A : X \to X$ and let $S_k = \{x : x \in X, A^kx = x\}$ for $k = 1,2,3,\ldots$; if, for some $m \geq 1$, $S_m = \{x_o\}$, then $S_1 = \{x_o\}$.
(This is trivial. Indeed, $A^mx_o = x_o \Rightarrow A^m(Ax_o) = Ax_o \Rightarrow Ax_o \in S_m \Rightarrow Ax_o = x_o \Rightarrow x_o \in S_1 \Rightarrow S_1 \neq \emptyset$; moreover, $\xi \in S_1 \Rightarrow \xi = A\xi = A^2\xi = \ldots = A^m\xi \Rightarrow \xi \in S_m \Rightarrow \xi = x_o \Rightarrow S_1 = \{x_o\}$).

Now, choose $y \in N_a$ and define $M_y : N_a \to N_a$ by

$$M_y x = (I+PZ)x-y.$$

Linearity of Z implies that, for any integer $k \geq 1$ and $x_1, x_2 \in N_a$,

$$M_y^k x_1 - M_y^k x_2 = (I+PZ)^k (x_1 - x_2). \qquad (*)$$

Thus, by our hypothesis, M_y^n is a contraction on N_a, and since N_a is a B-space, equation $M_y^n x = x$ has a unique solution in $N_a \Rightarrow M_y x = x$ has a unique solution in $N_a \Rightarrow PZx = y$ has a unique solution in $N_a \Rightarrow PZ$ is 1-to-1 from N_a onto N_a $\Rightarrow \mathfrak{h}$ is regular on $\mathcal{H}$ and $Q(\mathcal{H}) = \mathcal{H}$. The rest follows from Problem 1.3.1.

1.3.4 Let $Q_y : N_a \to N_a$ be defined by (1.3.9). If $y_1, y_2 \in N_a$, then $x_j = Q_{y_j} x_j \Leftrightarrow PZx_j = y_j$, $j = 1,2$ and we have by (1.3.9), (1.3.10), $\|x_1 - x_2\| \leq \|Q_{y_1} x_1 - Q_{y_1} x_2\| + \|Q_{y_1} x_2 - Q_{y_2} x_2\| \leq \mu \|x_1 - x_2\| + \|(PZ)^{-1}(y_1 - y_2)\|$, i.e.,

$$\|(PZ)^{-1} y_1 - (PZ)^{-1} y_2\| \leq (c-\lambda)^{-1} \|y_1 - y_2\|. \qquad (*)$$

Thus, for $e_1, e_2 \in \mathcal{H}$,

$$\|(PZ)^{-1} Pe_1 - (PZ)^{-1} Pe_2\| \leq (c-\lambda)^{-1} \|e_1 - e_2\|. \qquad (\ast\ast)$$

Now, define the operator S by $A = (PZ_1)^{-1}P+S$ with $A = (PZ)^{-1}P$, and denote $M = P(Z_1+Z_2)$, $N = PZ_1$; then $S = M^{-1}P - N^{-1}P = -N^{-1}(M-N)M^{-1}P$, and we have for $e_1, e_2 \in \mathcal{H}$ by $(\ast\ast)$,

$$\begin{aligned} \|Se_1 - Se_2\| &\leq c^{-1} \|(M-N)M^{-1}Pe_1 - (M-N)M^{-1}Pe_2\| \leq c^{-1}\lambda \|M^{-1}Pe_1 - M^{-1}Pe_2\| \\ &\leq c^{-1}\lambda(c-\lambda)^{-1}\|e_1 - e_2\| \end{aligned}$$

what we wished to show.

1.3.5 Due to Theorem 1.3.4, every $\mathfrak{h}_n$ is regular and $Q(\mathcal{H}) = \mathcal{H}$; by Lemma 1.3.1 it follows that $\|(PZ_n)^{-1}\| \leq \alpha^{-1}$ for each n. Choose $e \in \mathcal{H}$; then we have with $u = (PZ_o)^{-1}Pe$,

$$\begin{aligned} \|i_n - i_o\| &= \|(PZ_n)^{-1}Pe - (PZ_o)^{-1}Pe\| = \|-(PZ_n)^{-1}(P(Z_n - Z_o))(PZ_o)^{-1}Pe\| \\ &\leq \alpha^{-1}\|P\| \cdot \|Z_n u - Z_o u\| \to 0 \end{aligned}$$

as $n \to \infty$.

Section 1.4

1.4.1 (a) Let Z be bounded; then

$$\|M(x,y)\|^2 = \|Zx+a^*y\|^2 + \|ax\|^2 \le (\|Zx\| + \|a^*y\|)^2 + \|ax\|^2$$
$$\le 2\|Zx\|^2 + 2\|a^*y\|^2 + \|ax\|^2 \le \max[2\|Z\|^2 + \|a\|^2,\ 2\|a\|^2]\cdot\|(x,y)\|^2.$$

(b) Let M be bounded, i.e., there exists $\alpha > 0$ such that

$$\|Zx+a^*y\|^2 + \|ax\|^2 \le \alpha^2(\|x\|^2 + \|y\|^2)$$

for all $(x,y) \in \mathcal{H} \times N_{a^*}^{\perp}$. Putting $y = 0$, we get

$$\|Zx\|^2 + \|ax\|^2 \le \alpha^2\|x\|^2 \Rightarrow \|Zx\| \le \alpha\|x\|.$$

Section 1.6

1.6.1 Elementary.

1.6.2 Elementary.

1.6.3 Yes. Let $\mathcal{H} = R^2$, and for each $(x_1,x_2) \in R^2$ let

$$\begin{aligned} S_T(x_1,x_2) &= (0,0) \text{ for } T < -1 \\ &= (x_1,0) \text{ for } -1 \le T < 1 \\ &= (x_1,x_2) \text{ for } 1 \le T. \end{aligned}$$

Then $\{S_T : T \in R^1\}$ is a resolution of identity on R^2. Moreover, let $A : R^2 \to R^2$ be defined by $Ax = \begin{bmatrix} 1 & 1 \\ 1 & -1 \end{bmatrix}\cdot x$. Clearly, A is not causal, but $A^2 = 2I$ (I - identity operator) is causal.

1.6.4 Let A be defined by $(Ax)(t) = x(\frac{t}{2})$. Then A is causal, linear and bounded, since $\|Ax\| = \sqrt{2}\,\|x\|$ for each $x \in L_2[0,\infty)$. Moreover, the inverse A^{-1} exists and is defined by $(A^{-1}x)(t) = x(2t)$. Clearly, A^{-1} is not causal.

1.6.5 Since h is regular on D, PZ is simple on $N_a \cap D$. Choose $T \in R^1$ and $x_j \in N_a \cap D$ such that $S_T x_1 \ne S_T x_2$. By (ii), $S_T(x_1-x_2) = PS_T(S_T(x_1-x_2))$, and (i) yields

$$\langle S_T Pz_1 - S_T Pz_2, S_T x_1 - S_T x_2\rangle \ne 0.$$

Thus, if $S_T x_1 \ne S_T x_2$, we have for any $w_j \in S_T PZx_j$, $j = 1,2$,

$$\langle w_1 - w_2, S_T x_1 - S_T x_2\rangle \ne 0. \qquad (*)$$

Let $y_j \in [(PZ)(N_a \cap D)]^{\circ}$ be such that $S_T(PZ)^{-}y_1 \ne S_T(PZ)^{-}y_2$, and put $x_j = (PZ)^{-}y_j \in N_a \cap D$; then $S_T x_1 \ne S_T x_2$ and $y_j \in PZx_j$, i.e., $S_T y_j \in S_T PZx_j$, and $(*)$ gives

$$\langle S_T y_1 - S_T y_2, S_T(PZ)^{-}y_1 - S_T(PZ)^{-}y_2\rangle \ne 0. \qquad (\overset{*}{*})$$

However, $(\overset{*}{*})$ shows that $S_T y_1 - S_T y_2 \ne 0$. Hence, $S_T(PZ)^{-}y_1 \ne S_T(PZ)^{-}y_2 \Rightarrow$

$S_T y_1 \neq S_T y_2$, i.e., $(PZ)^-$ is causal on $[(PZ)(N_a \cap D)]^o$. The rest is obvious.

Section 2.1

2.1.1 Elementary

2.1.2 Define the operator $S : H^{c_o} \to H^{c_o}$ by

$$S = \hat{X}_1^* \hat{X}_2. \qquad (*)$$

By (i), (v) in Lemma 2.1.2, S is bounded; also, $SH^{c_o} = H^{c_o}$. Indeed, since $\hat{X}_2 H^{c_o} = N_{\hat{a}}$, we have $SH^{c_o} = \hat{X}_1^* \hat{X}_2 H^{c_o} = \hat{X}_1^* N_{\hat{a}}$; however, because $\hat{X}_1 H^{c_o} = N_{\hat{a}}$, it follows by (2.1.14) that $\hat{X}_1^* \hat{X}_1 H^{c_o} = H^{c_o} = \hat{X}_1^* N_{\hat{a}}$ which proves our claim.

Moreover, on H^{c_o} we have

$$\hat{X}_1 S = \hat{X}_2. \qquad (\ast\ast)$$

Indeed, let $x \in H^{c_o}$. Then $\hat{X}_1 Sx = \hat{X}_1 \hat{X}_1^* \hat{X}_2 x$; however, $\hat{X}_2 x \in N_{\hat{a}}$, and by Lemma 2.1.2 (vii), $\hat{X}_1 \hat{X}_1^*$ is the orthogonal projection onto $N_{\hat{a}}$. Hence, $(\hat{X}_1 \hat{X}_1^*)\hat{X}_2 x = \hat{X}_2 x$ which proves $(\ast\ast)$.

Next, assume that $Sx = 0$ for some $x \in H^{c_o}$; then, due to $(\ast\ast)$, $\hat{X}_2 x = 0$ $\Rightarrow x = 0$, i.e., S is 1-to-1.

Finally, by $(*)$, $\|S\| \leq \|\hat{X}_1^*\| \cdot \|\hat{X}_2\| = 1$, and by $(\ast\ast)$, $\|\hat{X}_2\| = 1 \leq \|\hat{X}_1\| \cdot \|S\| = \|S\|$; hence, $\|S\| = 1$.

2.1.3 The proof follows immediately from Lemma 2.1.2 by using the fact that $Y = X \cdot M$, where M is a nonsingular $c_o \times c_o$ matrix.

Section 2.2

2.2.1 Elementary application of Problem 2.1.3.

Section 2.4

2.4.1 Let $\tilde{A} : L_2[0,2\pi] \to L_2[0,2\pi]$ be defined by

$$(\tilde{A}x)(t) = \int_0^t x(\tau)d\tau. \qquad (*)$$

Then $\tilde{A}L_2 = K = \{y : y$ is absolutely continuous on $[0,2\pi]$, $y(0) = 0$, $y' \in L_2\}$; clearly K is not closed in L_2. For $n = 0, \pm1, \pm2, \ldots$, let $e_n = (2\pi)^{-\frac{1}{2}} e^{int}$; then the system $\{e_o, e_1, e_{-1}, e_2, e_{-2}, \ldots\}$ is an orthonormal basis in $L_2[0,2\pi]$. Moreover, define the operator T on $L_2[0,2\pi]$ by

$$Tx = [\langle x,e_o\rangle, \langle x,e_1\rangle, \langle x,e_{-1}\rangle, \langle x,e_2\rangle, \ldots]^T; \qquad (\ast\ast)$$

as known, T is an isometric isomorphism between $L_2[0,2\pi]$ and ℓ_2. By a routine calculation it follows from $(*)$ and $(\ast\ast)$ that, for any $x \in L_2[0,2\pi]$,

$T\tilde{A}x = A\cdot Tx$, where A is a symmetric $\aleph_0 \times \aleph_0$ matrix with $\pi, \frac{i}{1}, -\frac{i}{1}, \frac{i}{2}, -\frac{i}{2}, \ldots$ being the elements in the first row, $\pi, -\frac{i}{1}, \frac{i}{1}, -\frac{i}{2}, \frac{i}{2}, \ldots$ being the elements on the main diagonal, and with other elements being zero. Thus, A has the property that $\sum_{ik} |a_{ik}|^2 < \infty$. Now, if the operator $\hat{A} : \ell_2 \rightarrow \ell_2$ is defined by $\hat{A}\xi = A\cdot\xi$, we get $T\tilde{A}L_2 = \hat{A}TL_2$; since $TL_2 = \ell_2$, we have $TK = \hat{A}\ell_2$. However, because T is a homeomorphism and K is not closed, $\hat{A}\ell_2$ is not closed, either. This disproves our conjecture.

Section 3.5

3.5.1 (a) Suppose that $i = [i_1, i_2, i_3]^T \in D$ is a solution. By definition of D, $i_j(t)$, $j = 1,2,3$, is absolutely continuous on $[0,\tau]$ and $i_j(0) = 0$. Let $\gamma = [1,1,0]^T$; then $a\cdot\gamma = 0$ and we have by (2.2.2), $i_1 + 2i_2 + \int_0^t i_2 d\tau - 1 = 0$. This is an equality in $L_2[0,\tau]$, i.e., the equation is satisfied for almost every t in $[0,\tau]$. However, since $i_1 + 2i_2 + \int_0^t i_2 d\tau$ is a continuous function, it follows that the above equation holds for all $t \in [0,\tau]$. Hence, setting $t = 0$, we get $-1 = 0$ which is a contradiction.

(b) Let $i \in D$ be an elementary solution. If $\gamma \in R^{c_2}$ is such that $a\cdot\gamma = 0$, then (2.2.2) holds. Integrating this relation between limits 0,t, we obtain (2.2.2) where $\hat{Z}$ is replaced by $\hat{\hat{Z}}$, and e by $\int_0^t e(\tau)d\tau$. Hence, i is a generalized solution.

Section 4.3

4.3.1 Let $H = R^1$, let $D = R^3$ and let $\hat{Z} : R^3 \rightarrow R^3$ be defined by $\hat{Z}[x_1, x_2, x_3]^T = [0,1,1]$. Consider the network $\hat{n} = (\hat{Z}, G)$ whose graph G is given in Figure P.1

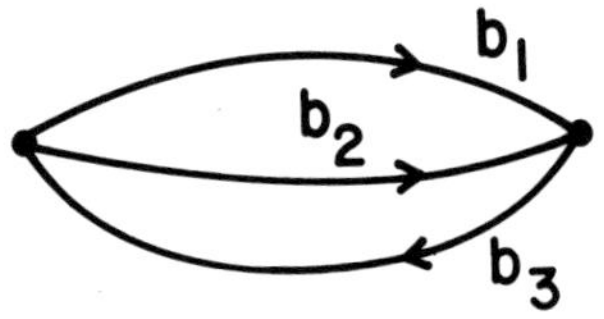

Figure P.1

Clearly, $F = R^2 \neq \phi$, but $Q_1 = \phi$. Indeed, suppose there exists $e \in R^1$ which is in Q_1. Then $\tilde{e} = [e,0,0]^T \in Q(D)$, and consequently, $\hat{n}$ possesses some solution $i = [i_1, i_2, i_3]^T \in R^3$ corresponding to $\tilde{e}$. On the other hand, the vector $\gamma = [0,1,1]^T$ satisfies the equation $a\cdot\gamma = 0$, and consequently, by

(2.2.2), $\bar{\gamma}^T \cdot (\hat{Z}i - \tilde{e}) = 0$. However, $\bar{\gamma}^T \cdot (\hat{Z}i - \tilde{e}) = 2$ which is a contradiction.

4.3.2 Inequality (*) implies by Lemma 1.3.1 (b) that W^{-1} exists and $W^{-1} \in [H^{c_o}, H^{c_o}]$. Moreover, putting $x = W^{-1}z$ in (*), we get

$$|\langle W^{-1}z, z\rangle_{c_o}| \geq \gamma \|W\|^{-2} \|z\|^2_{c_o}. \qquad (\ast\ast)$$

Thus, if A is the admittance of $\hat{\eta}$, we have for any $e \in H^{c_2}$ (note that $Q(D) = H^{c_2}$),

$$|\langle Ae, e\rangle_{c_2}| = |\langle \hat{X}W^{-1}\hat{X}^* e, e\rangle_{c_2}| = |\langle W^{-1}\hat{X}^* e, \hat{X}^* e\rangle_{c_o}| \qquad (\ast\ast\ast)$$
$$\geq \gamma \|W\|^{-2} \|\hat{X}^* e\|^2_{c_o}.$$

Now, the operator $A_n : Q_n = H^n \to D_n$ defined by $A_n x = (A(x)')_n$ is clearly in $[H^n, H^n]$, and $(\ast\ast\ast)$ yields by Lemma 4.1.1 for every $x \in H^n$,

$$|\langle A_n x, x\rangle_n| = |\langle (A(x)')_n, x\rangle_n| = |\langle A(x)', (x)'\rangle_{c_2}|$$
$$\geq \gamma \|W\|^{-2} \|\hat{X}^*(x)'\|^2_{c_o} \geq \gamma \|W\|^{-2} \lambda^2 \|x\|^2_n.$$

However, this inequality shows by Lemma 1.3.1 (b) that A_n^{-1} exists and is in $[H^n, H^n]$. Since $R = A_n^{-1}$, the proof is complete.

4.3.3 If $x \in Mj$, then $x \in Q_n$ and there exists a solution i_x of $\hat{\eta}$ corresponding to $(x)'$ such that $(i_x)_n = j$. Thus $x' \in Q(D)$ and $i_x \in A(x)'$, i.e., $j \in (A(x)')_n = A_n x$; hence, $x \in A_n^{-}j = Rj$. Consequently, $Mj \subset Rj$. The converse inclusion follows from the Proposition in Section 4.1.

Section 6.2

6.2.1 By (6.2.7), each element A'_{ik} of the matrix A' generating the admittance A of $\hat{\eta}$ is in $\mathfrak{U}$. On the other hand (6.2.1), (6.2.2), (6.2.3) show that each operator $M \in \mathfrak{U}$ is continuous, i.e., $Mx_n \to Mx$ in $\mathcal{D}'_+$ whenever $x_n \to x$ in $\mathcal{D}'_+$. The continuity of addition in $\mathcal{D}'_+$ concludes the proof.

REFERENCES

[1] G. F. Simmons, Introduction to topology and modern analysis, McGraw-Hill, N.Y., 1963.

[2] R. T. Rockafellar, On the maximality of sums of nonlinear monotone operators, Trans. Amer. Math. Soc., May 1970, pp. 75-88.

[3] S. K. Berberian, Introduction to Hilbert space, Oxford Univ. Press, N.Y., 1961.

[4] G. Bachman, L. Narici, Functional analysis, Acad. Press, N.Y., 1966, p. 275.

[5] R. Saeks, Causality in Hilbert space, SIAM Review, Vol. 12 (1970), pp. 357-383.

[6] C. A. Desoer, E. S. Kuh, Basic circuit theory, McGraw-Hill, N.Y., 1969.

[7] J. Jeans, Electricity and magnetism, 5th edition, Cambridge Univ. Press, London, 1927.

[8] V. Dolezal, A. H. Zemanian, Hilbert networks II: Some qualitative properties, SIAM J. Control, Vol. 13 1975) pp. 153-161.

[9] R. G. Bartle, The elements of integration, John Wiley, N.Y., 1966.

[10] C. E. Shannon, D. W. Hagelbarger, Concavity of resistance functions, J. Appl. Phys., 27 (1956) pp. 42-43.

[11] H. M. Melvin, On concavity of resisistance functions, Ibid, 27 (1956) pp. 658-659.

[12] I. W. Sandberg, A. N. Willson, Jr., Some theorems on properties of DC equations of nonlinear networks, SIAM J. Appl. Math., 22 (1972) pp. 173-186.

[13] V. Dolezal, Dynamics of linear systems, Academia, Prague/Noordhoff Groningen, 1967.

[14] A. H. Zemanian, Distribution theory and transform analysis, McGraw-Hill, N.Y., 1965.

[15] V. Dolezal, The existence of a continuous basis of a certain linear subspace of E_r which depends on a parameter, Cas. pest. matemat., 89 (1964), Prague, pp. 466-469.

[16] ________, Zur Dynamik der Linearsysteme, Acta Technica, 1960, pp. 19-33.

[17] J. L. Synge, The fundamental theorem of electric networks, Quarterly of Appl. Math., vol. 9, No. 2, July 1951.

Subject Index